ADVANCED MATHS FOR AQA
Statistics
Graham Upton Ian Cook
Course consultant: John White
Coursework guidance: Craig Simms
S2
D0525769
WITHDRAWN

OXFORD
UNIVERSITY PRESS

OXFORD
UNIVERSITY PRESS

Great Clarendon Street, Oxford OX2 6DP

Oxford University Press is a department of the University of Oxford.
It furthers the University's objective of excellence in research, scholarship,
and education by publishing worldwide in

Oxford New York

Auckland Cape Town Dar es Salaam Hong Kong Karachi
Kuala Lumpur Madrid Melbourne Mexico City Nairobi
New Delhi Shanghai Taipei Toronto

With offices in

Argentina Austria Brazil Chile Czech Republic France Greece
Guatemala Hungary Italy Japan South Korea Poland Portugal
Singapore Switzerland Thailand Turkey Ukraine Vietnam

Oxford is a registered trade mark of Oxford University Press
in the UK and in certain other countries

First published 2005

British Library Cataloguing in Publication Data

Data available

ISBN 0 19 914986 0

1 3 5 7 9 10 8 6 4 2

Typeset by Tech-Set Ltd, Gateshead, Tyne and Wear
Printed and bound in Great Britain by Bell and Bain.

Acknowledgements
The publishers would like to thank AQA for their kind permission to reproduce past
paper questions. AQA accept no responsibility for the answers to the past paper
questions which are the sole responsibility of the publishers.

The publishers would also like to thank James Nicholson for his authoritative guidance
in preparing this book.

The photograph on the cover is reproduced courtesy of Photodisc.

About this book

This Advanced level book is designed to help you get your best possible grade in the AQA S2 module in GCE Mathematics (MS2A and MS2B) for first examination in 2005. This module can contribute to an award in GCE A-level Mathematics.

Important note:
This module is different to the second module in GCE Statistics (SS02).

Each chapter starts with an overview of what you are going to learn and a list of what you should already know. The 'Before you start' section contains 'Check in' questions, which will help to prepare you for the topics in the chapter.

You should know how to ...	Check in
1 Summarise data using classes.	**1** Summarize the following data, using the classes 0–, 10–, 20–, 30–50: 5 15 23 34 4 32 45 8 18 22 22 44 49 21 16 18 19 2 26 31

Key information is highlighted in the text so you can see the facts you need to learn. Also hint boxes show tips and reminders you may find useful.

$$F(x) = P(X \leqslant x) = P(X < x) \qquad (3.2)$$

The function F is also called the **cumulative distribution function**, abbreviated to **cdf**.

Worked examples showing the key skills and techniques you need to develop are shown in boxes.

Example 4

Determine the cumulative distribution function for the random variable X defined as 'the result of rolling a fair six-faced die'.

Consider, for example, $P(X \leqslant 3)$. This is given by:

$$P(X \leqslant 3) = P(X = 1) + P(X = 2) + P(X = 3) = \frac{1}{6} + \frac{1}{6} + \frac{1}{6}$$
$$= \frac{3}{6}$$

The cumulative distribution function is therefore:

x	<1	1–	2–	3–	4–	5–	≥6
$P(X \leqslant x)$	0	$\frac{1}{6}$	$\frac{2}{6}$	$\frac{3}{6}$	$\frac{4}{6}$	$\frac{5}{6}$	$\frac{6}{6} = 1$

'Calculator practice' boxes suggest ways in which you can use your calculator in your work.

Calculator practice

If you have a graphical calculator then you should use it to check your sketch of the graph of F for $1 \leqslant x \leqslant 2$ and $2 \leqslant x \leqslant 4$.

The questions in the exercises are carefully selected to provide basic practice at the beginning, and harder questions towards the end.

At the end of each chapter there is a summary. The 'You should now be able to' section is useful as a quick revision guide, and each 'Check out' question identifies important techniques that you should remember.

You should now be able to ...	Check out
1 Use tables of the t-distribution.	**1** a) If X has a t_6-distribution, find c, where $P(X < c) = 0.99$. b) If X has a t_{12}-distribution, find c, where $P(X > c) = 0.05$.

Following the summary you will find a revision exercise with past paper questions from AQA. These will enable you to become familiar with the style of questions you will see in the exam.

Towards the end of the book there are two Practice Papers, one for Option A (with coursework) and one for Option B (without coursework). These will directly help you to prepare for your exams.

The last chapter is entitled 'Coursework guidance', and is for students taking the MS2A unit (with coursework). The chapter tells you what you have to do for your coursework, and contains advice from a senior moderator to help you improve your grade.

At the end of the book you will find numerical answers, statistical tables, a list of formulae you need to become familiar with, and a list of useful mathematical notation.

Contents

1 Discrete random variables

This chapter will show you how to

- ✦ Calculate the mean, variance and standard deviation of a discrete random variable
- ✦ Calculate the mean, variance and standard deviation of simple functions of a random variable

A brief introduction to discrete random variables appeared in Chapter 3 of *Statistics S1*.

Before you start

You should know how to ...	Check in
1 Identify sets of exhaustive events.	**1** A six-faced die is rolled. The events A, B, C and D are defined as follows. A: An even number is obtained. B: A multiple of 3 is obtained. C: A prime number (2, 3, or 5) is obtained. D: An odd number is obtained. Determine which pairs of events are exhaustive.
2 Identify sets of mutually exclusive events.	**2** For the events in question **1**, determine which pairs are mutually exclusive.
3 Apply probability laws.	**3** Determine the probability that: a) when a fair die is rolled, it shows either a 4 or a 5 b) when two fair dice are rolled, a 4 and a 5 are obtained c) when a fair die is rolled twice, the number obtained on the second roll is greater than that obtained on the first roll.
4 Calculate a relative frequency.	**4** Calculate the relative frequency of the number 4 in the sample of observations: 2, 4, 4, 6, 5, 9, 3.
5 Calculate a population mean, μ.	**5** Calculate the mean of the following population: 3, 7, 5, 12, 10, 8, 7, 7, 5, 12.
6 Calculate a population variance, σ^2, (using the divisor n).	**6** Calculate the variance of the population in question **5**.

1.1 Discrete random variables and their probability distributions

This chapter is concerned with discrete random variables. Recall that:

- A **variable** is a characteristic measured or observed when an experiment or trial is carried out or an observation is made.
- When the value of a variable may be subject to random variation, the variable is described as being a **random** variable.
- A random variable is described as being **discrete** when (in theory) a list can be made of its possible (numerical) values.

A simple example is the number on the number plate of the next car that passes a given point.

Notation

Random variables are usually denoted by italic capital letters, such as X, Y, Z. Observed values of these random variables are usually denoted by corresponding italic lower-case letters, such as x, y, z.

This leads to a statement such as:

$$P(X = x) = \frac{1}{4}$$

which should be read as:

'The probability that the random variable X takes the value x is $\frac{1}{4}$'

Probability distributions

Suppose a die with faces numbered 1 to 6 is rolled. Define the random variable X to be 'the number showing on the top of the die'. Two facts are known:

1. The observed value of X must be 1, 2, 3, 4, 5, or 6.
2. On a given roll, the random variable X can only take *one* of those values.

These two facts correspond to statements that the six outcomes are both collectively exhaustive and mutually exclusive. Hence:

The six outcomes are exhaustive because there are no other possible outcomes. They are mutually exclusive because if one occurs then the others can not.

$$P(X = 1) + P(X = 2) + \cdots + P(X = 6) = 1$$

Generalising, for a discrete random variable X, which can take only the distinct values $x_1, x_2, \ldots, x_n$, the result is:

$$P(X = x_1) + P(X = x_2) + \cdots + P(X = x_n) = 1 \qquad (1.1)$$

The notation can be simplified by writing $P(X = x_i)$ as p_i, so that:

$$\sum p_i = 1$$

The sizes of $p_1, p_2, \ldots, p_n$ show how the total probability is *distributed* amongst the possible values of X. The values of $p_1, p_2, \ldots, p_n$, together with the values $x_1, x_2, \ldots, x_n$ to which they refer, are said to define a **probability distribution**.

This is analogous to a frequency distribution.

A probability distribution is often displayed as a table of the possible values and their probabilities.

Example 1

Tabulate the probability distribution of the number of Heads obtained when a fair coin is tossed twice.

Let X be the random variable which denotes the number of Heads obtained. The possible values are 0, 1 and 2. The simplest way of finding the required probabilities is to use a probability tree diagram.

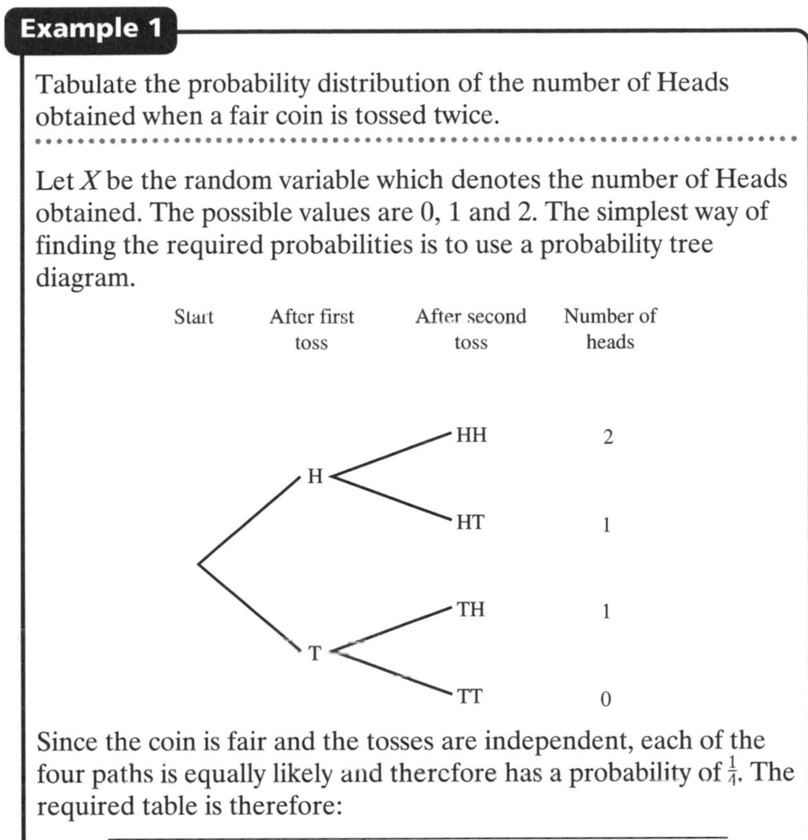

Since the coin is fair and the tosses are independent, each of the four paths is equally likely and therefore has a probability of $\frac{1}{4}$. The required table is therefore:

Number of Heads, x	0	1	2	All other values
$P(X = x)$	$\frac{1}{4}$	$\frac{2}{4} = \frac{1}{2}$	$\frac{1}{4}$	0

S2

The final column in this table is a statement (often omitted) that no values other than 0, 1 or 2 are possible (this is implied by the fact that the sum of the probabilities for $x = 0, 1, 2$ is one).

The probability function

For many situations there is no need to make a list of all the probabilities, because some simple all-embracing formula (sometimes called the **probability function**) can be found.

Example 2

Obtain a formula for the probability distribution of the random variable X defined as: 'The result of rolling a fair six-faced die'.

Each of the six possible values for X has a probability of $\frac{1}{6}$, so:

$$P(X = x) = \begin{cases} \frac{1}{6} & x = 1, 2, \ldots, 6 \\ 0 & \text{otherwise} \end{cases}$$

Illustrating probability distributions

As always in Statistics, it is a good idea to draw pictures whenever possible. Since a discrete random variable can only take discrete values, a bar chart is appropriate, with the vertical axis measuring probability.

Example 3

The random variable X is defined as: 'The sum of the numbers shown by two fair six-faced dice'. Tabulate the probability distribution of X and represent it on an appropriate diagram.

There are 36 possible outcomes, all of which are equally likely (since the dice are fair and the tosses are independent). These outcomes are illustrated in the table in which the entries are the sums of the numbers shown by the two dice.

Second die \ First die	1	2	3	4	5	6
1	2	3	4	5	6	7
2	3	4	5	6	7	8
3	4	5	6	7	8	9
4	5	6	7	8	9	10
5	6	7	8	9	10	11
6	7	8	9	10	11	12

The table shows that there is just 1 outcome for which $X = 2$. Therefore:

$$P(X = 2) = \frac{1}{36}$$

The full distribution is tabulated below:

x	2	3	4	5	6	7	8	9	10	11	12
$P(X = x)$	$\frac{1}{36}$	$\frac{2}{36}$	$\frac{3}{36}$	$\frac{4}{36}$	$\frac{5}{36}$	$\frac{6}{36}$	$\frac{5}{36}$	$\frac{4}{36}$	$\frac{3}{36}$	$\frac{2}{36}$	$\frac{1}{36}$

The distribution is illustrated in the following bar chart.

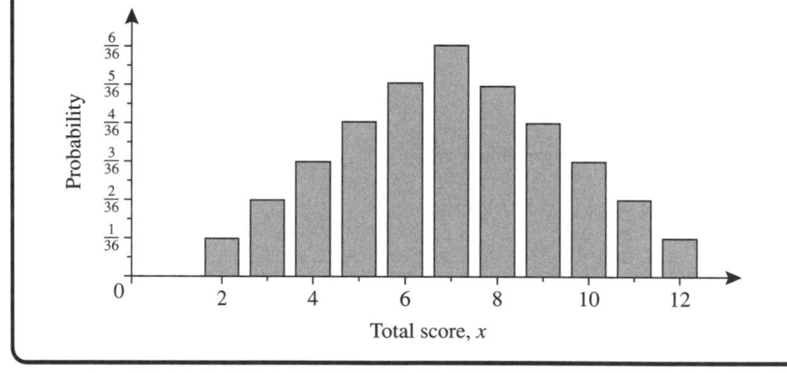

The most likely value for X is the value with the highest probability (the **mode**). In this case the mode is 7.

Exercise 1A

For the following questions, use a probability tree diagram to find, in each case, the set of possible values of the random variable X, and draw up a table showing $P(X = x)$ for each value of x.

1 With equal probability, Adam chooses between the numbers 0 and 1. Ben chooses between the numbers 8 and 9 and is twice as likely to choose 8 as 9. The random variable X is the total of the two numbers chosen.

2 A box contains three red marbles, four blue marbles, and five green marbles. Two marbles are taken at random *without* replacement. The random variable X is the number of green marbles obtained.

> 'Without replacement' means that after a marble is taken from the box it is not replaced in the box.

3 A box contains three red marbles, four blue marbles, and five green marbles. Two marbles are taken at random *with* replacement. The random variable X is the number of green marbles obtained.

> 'With replacement' means that after a marble is taken from the box it is put back in the box before the next marble is taken from the box.

4 A gambler is equally likely to lose £1 or gain £2. If he loses then he does not play again. If he wins then he plays for one final time, again being equally likely to lose £1 or to gain £2. Display the situation using a probability tree diagram and determine the probability distribution of X, the amount (in £) by which his fortune changes as a result of the above actions.

The cumulative distribution function

This is an alternative function for summarising a probability distribution. The function F is defined by

$$\mathrm{F}(x) = \mathrm{P}(X \leqslant x) = \sum_{x_i \leqslant x} \mathrm{P}(X = x_i) \qquad (1.2)$$

The function is defined for all values of x.

F(x) = 0 ---------- Smallest value —— Possible values of X —— Largest value ---------- F(x) = 1

It is equal to 0 for all values of x smaller than the smallest of the possible values of X, and it is equal to 1 for all values of x larger than the largest of the possible values of X.

Example 4

Determine the cumulative distribution function for the random variable X defined as 'the result of rolling a fair six-faced die'.

Consider, for example, $\mathrm{P}(X \leqslant 3)$. This is given by:

$$\mathrm{P}(X \leqslant 3) = \mathrm{P}(X = 1) + \mathrm{P}(X = 2) + \mathrm{P}(X = 3) = \frac{1}{6} + \frac{1}{6} + \frac{1}{6}$$
$$= \frac{3}{6}$$

The cumulative distribution function is therefore:

x	<1	1–	2–	3–	4–	5–	$\geqslant 6$
$\mathrm{P}(X \leqslant x)$	0	$\frac{1}{6}$	$\frac{2}{6}$	$\frac{3}{6}$	$\frac{4}{6}$	$\frac{5}{6}$	$\frac{6}{6} = 1$

The graph of the cumulative distribution function for any discrete random variable looks like a flight of steps. In the case of Example 4 the steps are particularly regular.

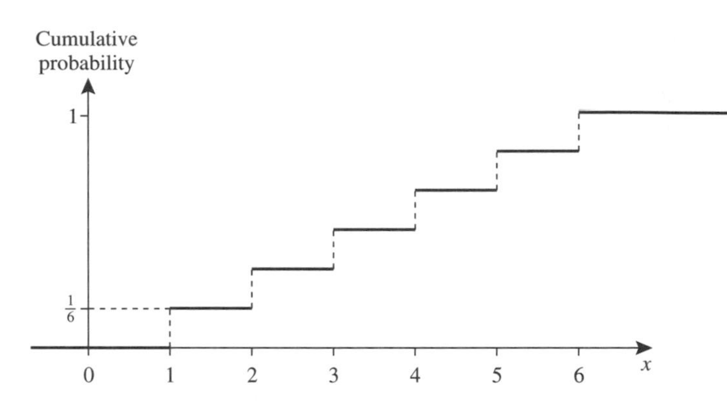

The discrete uniform distribution

This is the simplest of all the possible discrete distributions, since it prescribes that all the possible outcomes are equally likely to occur. The most familiar case is the rolling of a fair die where there are six possible and equally likely outcomes (see Example 4).

The situation can be generalised to the case of a random variable X that is equally likely to take any of k distinct values $x_1, x_2, \ldots, x_k$:

$$P(X = x) = \begin{cases} c & x = x_1, x_2, \ldots, x_k \\ 0 & \text{otherwise} \end{cases}$$

where c is some constant to be determined.

Now:

$$P(X = x_1) + P(X = x_2) + \cdots + P(X = x_k) = 1$$

so that:

$$c + c + \cdots + c = 1$$

which implies that $kc = 1$ and hence that $c = \dfrac{1}{k}$. The distribution is therefore properly specified by:

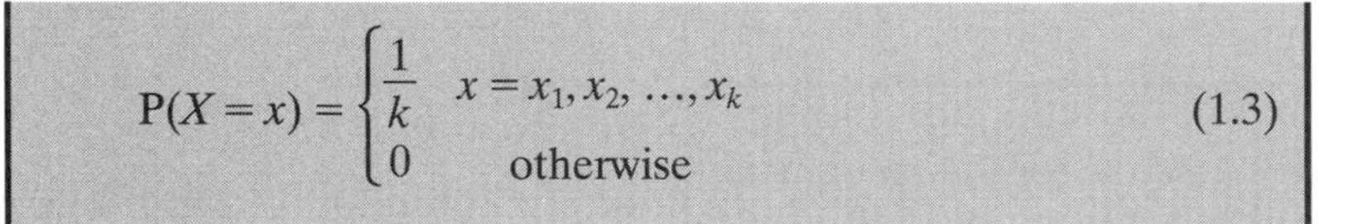

$$P(X = x) = \begin{cases} \dfrac{1}{k} & x = x_1, x_2, \ldots, x_k \\ 0 & \text{otherwise} \end{cases} \qquad (1.3)$$

Example 5

Write down the probability distribution of X, the random variable denoting the outcome of rolling a fair six-sided die.

The distribution is obviously:

x	1	2	3	4	5	6
$P(X = x)$	$\frac{1}{6}$	$\frac{1}{6}$	$\frac{1}{6}$	$\frac{1}{6}$	$\frac{1}{6}$	$\frac{1}{6}$

More compactly:

$$P(X = x) = \begin{cases} \frac{1}{6} & x = 1, 2, \ldots, 6 \\ 0 & \text{otherwise} \end{cases}$$

The diagram is illustrated by the following bar chart:

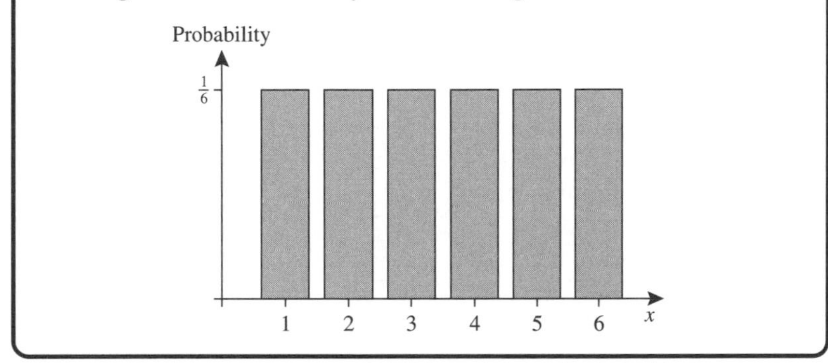

The corresponding cumulative distribution function was given in Example 4.

S2

The Bernoulli distribution

This is the distribution of a random variable X that can take only the values 0 and 1. With $P(X = 1)$ being denoted by p, where $0 \leqslant p \leqslant 1$, the distribution is specified by:

The distribution is named after the Swiss mathematician James Bernoulli (1654–1705).

$$P(X = x) = \begin{cases} 1 - p & x = 0 \\ p & x = 1 \\ 0 & \text{otherwise} \end{cases} \qquad (1.4)$$

An example of a random variable X with a Bernoulli distribution is: 'The number of Heads obtained on a single toss of a bent coin, for which the probability of a Head is p'.

The binomial distribution

An example of a binomial random variable would be: 'The total number of Heads obtained in a sequence of n independent tosses of a bent coin', where, for each toss, the probability of a Head is p. In this case:

The binomial distribution was discussed at length in Chapter 3 of *Statistics S1*. The requirement is that each of n independent trials has an identical probability, p, of being a 'success'; the random variable is the number of successes.

$$P(X = x) = \begin{cases} \binom{n}{x} p^x (1 - p)^{n - x} & x = 0, 1, \ldots, n \\ 0 & \text{otherwise} \end{cases}$$

The case $n = 1$ is the Bernoulli distribution.

Exercise 1B

1 A book has pages numbered 1 to 300. A page is chosen at random and X is the last digit of the page number. Find the probability distribution of X.

2 Determine the distribution of X, where X is the random variable denoting the number of sixes obtained on a single throw of a fair die.

3 The random variable X has a distribution which is both uniform and Bernoulli.

Write down the probability function of X.

4 The random variable X is the number of Heads obtained when a fair coin is thrown twice.

Show that the distribution of X is:

a) not uniform b) not Bernoulli.

5 Forty percent of the students in a large school are male, and the rest are female. A random sample of ten students is selected to represent the school. The random variable X is the number of females selected. State the distribution of X.

6 The random variables X, Y and Z are defined as follows:

X: The number of girls in a selection of 6 pupils, chosen at random from a class of 15 girls and 12 boys.

Y: The number of boys chosen, when one pupil is chosen at random from each of the thirty classes in a mixed-sex school.

Z: The number of candidates who obtain grade A in Mathematics out of a random sample of six candidates chosen from all those entered for the subject. It is known that 19% of all candidates entered for Mathematics obtained a grade A.

For each of the random variables, state whether or not a binomial distribution is a suitable model. If you think a binomial distribution is suitable, give appropriate values for the parameters. If you think a binomial distribution is not suitable, give one reason why it is not.

7 A book has pages numbered 1 to 300. A page is chosen at random and X is the first digit of the page number.

a) Does X have a discrete uniform distribution?

b) Tabulate the distribution of X.

S2

Probability distributions and relative frequency distributions

Consider, as an example, the values obtained in successive rolls of a fair six-faced die. Relative frequencies of the possible outcomes are recorded in the following table:

Each roll of the die is a **trial**.

Number of rolls	Outcome 1	2	3	4	5	6
36	0.222	0.083	0.167	0.167	0.139	0.222
216	0.185	0.139	0.167	0.162	0.185	0.162
1296	0.156	0.171	0.168	0.159	0.176	0.170
7776	0.162	0.164	0.173	0.160	0.168	0.173
46 656	0.168	0.164	0.167	0.168	0.165	0.168
Target	$\frac{1}{6}$	$\frac{1}{6}$	$\frac{1}{6}$	$\frac{1}{6}$	$\frac{1}{6}$	$\frac{1}{6}$

The relation between relative frequency and probability was introduced in Chapter 2 of *Statistics S1*.

Notice that the values of each of the relative frequencies become less variable as the sample size increases.

If you concentrate on a particular outcome, such as obtaining a six, and plot relative frequency against number of rolls, then you get a graph such as the following:

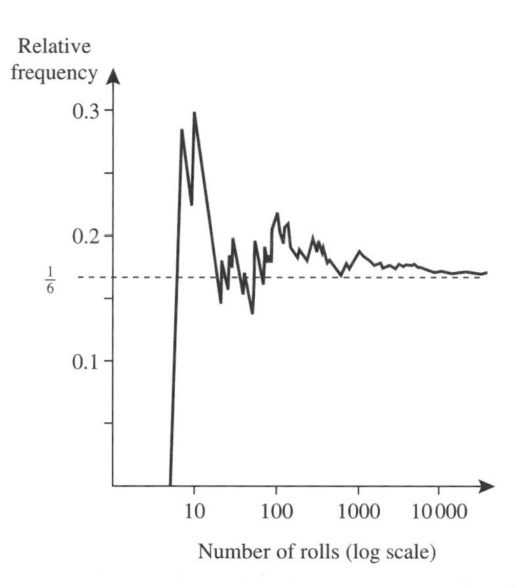

> As the number of trials increases the relative frequency of an event converges on its limiting value, the probability of the event.

As the sample size (the number of rolls) increases, so the relative frequencies tend to converge ever more closely on the theoretical probabilities, and the observed relative frequency distribution of the possible outcomes converges on the theoretical probability distribution.

> A probability distribution may be thought of as the limit of a relative frequency distribution as the sample size increases.

Calculator practice

Write your own computer (or calculator) program to simulate the tossing of a die. To convert a random number r having a value between 0 and 1 into the value of a die toss, x, set

$$x = 1 + \text{INT}(6 * r)$$

where INT is a function that truncates a decimal to an integer (for example, INT(5.8) = 5). Examine the relative frequencies as you increase the sample size – if they are all converging on $\frac{1}{6}$ then the program works!

S2

1.2 Mean, variance and standard deviation

The mean

The sample mean for data summarized as a frequency distribution is given by:

$$\bar{x} = \frac{\sum fx}{\sum f} = \frac{1}{n}\sum fx$$

Formulae for the calculation of the sample mean, variance (s^2) and standard deviation (s) were presented in Chapter 1 of *Statistics S1*.

where f is the frequency with which the value x occurs and n ($= \sum f$) is the total number of observations. The formula can be rewritten as:

$$\bar{x} = \sum\left\{x\left(\frac{f}{n}\right)\right\}$$

S2

which emphasises that, in the summation, each value of x is multiplied by its relative frequency.

An idea of what happens to $\bar{x}$ as the sample size increases, is obtained by re-examining the dice-rolling results (page 8) and plotting the value of the sample mean, $\bar{x}$, against the number of rolls:

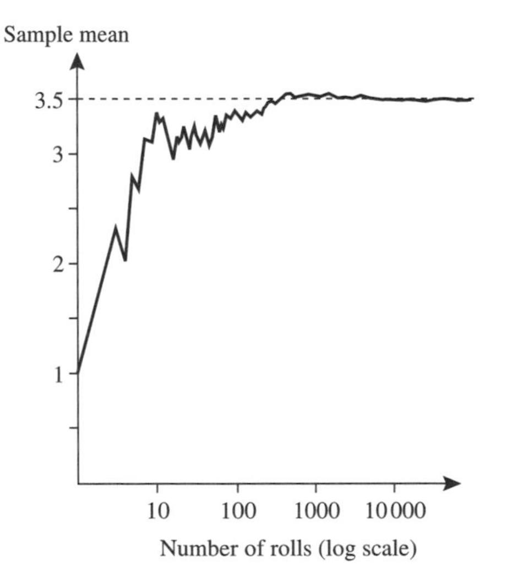

You can see that, as in the case of a relative frequency, after some initial oscillations the sample mean appears to be settling down to some value. This limiting value is the mean of X and is often referred to as the **expectation** or **expected value** of X, written as $E(X)$.

Since the limiting value of a relative frequency is a probability, the formula for the mean becomes:

$$E(X) = \sum x P(X = x) \qquad (1.5)$$

where the summation is over all possible values of x.

The mean of X, $E(X)$, is often denoted by μ. Hence:

$$\mu = E(X) = \sum_i x_i p_i$$

where $p_i = P(X = x_i)$ and the sum is over all values of i.

Example 6

Determine the expectation of the random variable X, which has probability distribution given below.

x	0	1	2	3
$P(X = x)$	0.3	0.4	0.2	0.1

$$\begin{aligned} E(X) &= (0 \times 0.3) + (1 \times 0.4) + (2 \times 0.2) + (3 \times 0.1) \\ &= 0 + 0.4 + 0.4 + 0.3 \\ &= 1.1 \end{aligned}$$

The expectation of X is 1.1.

$E(X)$ need not be an integer, nor does it have to be one of the possible values for X.

S2

Example 7

The random variable X can only take the values 2 and 5. Given that the value 5 is twice as likely as the value 2, determine the mean of X.

Denote $P(X = 2)$ by p. Then $P(X = 5) = 2p$. Since 2 and 5 are the only possible values for X, the sum of their probabilities is 1: $p + 2p = 1$. So $3p = 1$ and it follows that $p = \frac{1}{3}$. The mean of X is therefore given by:

$$\begin{aligned} E(X) = (2 \times P(X = 2)) + (5 \times P(X = 5)) &= \left(2 \times \frac{1}{3}\right) + \left(5 \times \frac{2}{3}\right) \\ &= \frac{2}{3} + \frac{10}{3} \\ &= 4 \end{aligned}$$

The random variable X has mean 4, though no individual values of X can be equal to that value.

Example 8

In a multiple choice paper, each question is followed by four alternative answers. The candidate is asked to ring one of these answers. If the answer ringed is correct, then the candidate gains 3 marks, but if the answer is incorrect the candidate loses 1 mark.

Determine the expected value of the mark gained for a question if:

a) the candidate chooses an answer at random,

b) the candidate knows that one of the incorrect answers is incorrect and chooses at random from the remaining three possibilities.

Comment on the results in each case.

Let X be the number of marks gained.

a) The probability distribution for X is:

$$P(X=3)=\frac{1}{4} \qquad P(X=-1)=\frac{3}{4}$$

so that:

$$E(X)=\left(3\times\frac{1}{4}\right)+\left(\{-1\}\times\frac{3}{4}\right)=\frac{3}{4}-\frac{3}{4}=0$$

The examination marking scheme has been designed so that the expected mark obtained by someone who knows nothing and guesses the answer at random will be zero.

b) The revised probability distribution, after the elimination of one of the incorrect answers is:

$$P(X=3)=\frac{1}{3} \qquad P(X=-1)=\frac{2}{3}$$

so that now:

$$E(X)=\left(3\times\frac{1}{3}\right)+\left(\{-1\}\times\frac{2}{3}\right)=1-\frac{2}{3}=\frac{1}{3}$$

Since $E(X)$ is greater than 0, it follows that if one or more of the possibilities can be eliminated as being certainly incorrect then there will be an advantage (averaging one-third of a mark) in guessing the answer at random.

S2

Expectation of X^2

Just as $E(X)$ is the average value of X calculated using probabilities, so $E(X^2)$ is the average value of X^2 calculated using probabilities:

$$E(X^2)=\sum x^2 P(X=x) \qquad (1.6)$$

where the summation is over all possible values of X. An alternative form is

$$E(X^2)=\sum_i x_i^2 p_i$$

where $p_i=P(X=x_i)$ and the summation is over all values of i.

Note that it is only the values of X that are squared, and not the probabilities.

Example 9

Calculate the expected value of X^2, where X is the value obtained from rolling a fair six-faced die.

Each face has probability $\frac{1}{6}$, hence:

$$\begin{aligned} E(X^2) &= \left(1^2 \times \frac{1}{6}\right) + \left(2^2 \times \frac{1}{6}\right) + \cdots + \left(6^2 \times \frac{1}{6}\right) \\ &= \frac{1}{6} + \frac{4}{6} + \cdots + \frac{36}{6} \\ &= \frac{91}{6} \end{aligned}$$

The expected value of X^2 is about 15 and is *not* simply the square of $E(X)$ (which would have given the answer $3.5^2 \approx 12$).

$E(X^2)$ is never less than $(E(X))^2$.

S2

Exercise 1C

1 The random variable X has probability distribution given by:

x	1	2	3	4
$P(X=x)$	0.2	0.4	0.1	0.3

a) Determine the values of $E(X)$ and $E(X^2)$.

b) Verify that $E(X^2) > (E(X))^2$.

2 The random variable X is equally likely to take the values 4, 5 and 6. Determine the values of $E(X)$ and $E(X^2)$.

3 The random variable Y has probability distribution given by:

$$P(Y=y) = \begin{cases} \frac{1}{9}y & y = 2, 3, \text{ or } 4 \\ 0 & \text{otherwise} \end{cases}$$

Determine the values of $E(Y)$ and $E(Y^2)$.

4 The random variable X has probability distribution given by:

$$P(X=x) = \begin{cases} kx & x = 1, 2, \text{ or } 3 \\ 0 & \text{otherwise} \end{cases}$$

a) Find the value of the constant k.

b) Determine the values of $E(X)$ and $E(X^2)$.

5 The random variable Y has probability distribution given by:

$$P(Y=y) = \begin{cases} ky^2 & y = 1, 2 \text{ or } 3 \\ 0 & \text{otherwise} \end{cases}$$

a) Find the value of the constant k and determine the values of $E(Y)$ and $E(Y^2)$.

b) Verify that $E(Y^2) > (E(Y))^2$.

The variance

As the sample size increases so the values of the sample statistics approach the values of the corresponding population statistics. In particular:

Sample statistic		Population statistic
Relative frequency, $\frac{f}{n}$	$\rightarrow$	Probability, p
Mean, $\bar{x}$	$\rightarrow$	Mean, $\mathrm{E}(X) = \mu$

and, in addition:

Variance, s^2	$\rightarrow$	Variance, $\mathrm{Var}(X)$

S2

The variance of a random variable is usually denoted as σ^2, where σ is the lower-case Greek letter corresponding to Σ and is therefore also called 'sigma'.

In terms of expectations, the variance is given by:

$$\sigma^2 = \mathrm{Var}(X) = \mathrm{E}(X^2) - (\mathrm{E}(X))^2 = \mathrm{E}(X^2) - \mu^2 \quad (1.7)$$

Thus $\mathrm{E}(X^2) = (\mathrm{E}(X))^2$ if, and only if, $\mathrm{Var}(X) = 0$.
The minimum value for σ^2 is, in fact, zero; this occurs when the random variable has only one possible value.

Example 10

The random variable X has probability distribution given by:

x	2	5
$\mathrm{P}(X = x)$	0.4	0.6

Determine the variance of X.

The expectation of X is given by:

$$\mathrm{E}(X) = (2 \times 0.4) + (5 \times 0.6) = 3.8$$

Similarly, $\mathrm{E}(X^2)$ is given by:

$$\mathrm{E}(X^2) = (2^2 \times 0.4) + (5^2 \times 0.6) = 16.6$$

Thus:

$$\mathrm{Var}(X) = \mathrm{E}(X^2) - (\mathrm{E}(X))^2 = 16.6 - 3.8^2 = 2.16$$

Example 11

The random variable X has probability distribution given by:

x	2	3	4
$\mathrm{P}(X = x)$	p	p	$1 - 2p$

a) Find the range of possible values for p.
b) Show that $\mathrm{Var}(X) = p(5 - 9p)$.

a) Since probabilities cannot be negative, $0 \leqslant p$ and also $0 \leqslant (1 - 2p)$. Adding $2p$ to each side of the latter inequality gives $2p \leqslant 1$. The range of possible values is therefore given by

$$0 \leqslant p \leqslant \frac{1}{2}$$

The range of values for p ensures that, for all x, $0 \leqslant \mathrm{P}(X = x) \leqslant 1$.

b) The expectation of X is given by:

$$\mathrm{E}(X) = (2 \times p) + (3 \times p) + (4 \times (1 - 2p)) = 4 - 3p$$

Similarly, $\mathrm{E}(X^2)$ is given by:

$$\mathrm{E}(X^2) = (2^2 \times p) + (3^2 \times p) + (4^2 \times (1 - 2p)) = 16 - 19p$$

Hence:

$$\begin{aligned}\mathrm{Var}(X) &= \mathrm{E}(X^2) - (\mathrm{E}(X))^2 \\ &= (16 - 19p) - (4 - 3p)^2 = 5p - 9p^2 = p(5 - 9p)\end{aligned}$$

as required.

S2

Example 12

The random variable X has the Bernoulli distribution:

$$\mathrm{P}(X = x) = \begin{cases} 1 - p & x = 0 \\ p & x = 1 \\ 0 & \text{otherwise} \end{cases}$$

where $0 \leqslant p \leqslant 1$.
Find the expectation and variance of X.

Finding $\mathrm{E}(X)$ is straightforward:

$$\mathrm{E}(X) = (0 \times (1 - p)) + (1 \times p) = p$$

Similarly:

$$\mathrm{E}(X^2) = (0^2 \times (1 - p)) + (1^2 \times p) = p$$

Hence:

$$\mathrm{Var}(X) = \mathrm{E}(X^2) - (\mathrm{E}(X))^2 = p - p^2 = p(1 - p)$$

The standard deviation

The standard deviation of the random variable X is the square root of the variance of X, and is denoted by σ.

Example 13

The discrete random variable X has probability distribution given by:

$$P(X = x) = \begin{cases} kx^3 & x = 1, 2, 3 \\ 0 & \text{otherwise} \end{cases}$$

Determine, correct to 3 significant figures, the values of

a) the constant k,
b) the mean of X,
c) the standard deviation of X.

a) Since the probabilities sum to 1:

$$k(1^3 + 2^3 + 3^3) = 1$$

The left-hand side is $36k$ and hence $k = \frac{1}{36} = 0.0272$ (to 3 sf).

b) The mean is $E(X)$, which is given by:

$$\begin{aligned} E(X) &= (1 \times k) + (2 \times 8k) + (3 \times 27k) \\ &= (1 + 16 + 81)k \\ &= 98k = \frac{98}{36} = \frac{49}{18} = 2.72 \text{ (to 3 sf)} \end{aligned}$$

c) Since:

$$\begin{aligned} E(X^2) &= (1^2 \times k) + (2^2 \times 8k) + (3^2 \times 27k) \\ &= (1 + 32 + 243)k \\ &= 276k = \frac{276}{36} = \frac{23}{3} \end{aligned}$$

it follows that:

$$\text{Var}(X) = E(X^2) - (E(X))^2 = \frac{23}{3} - \left(\frac{49}{18}\right)^2 = \frac{83}{324}$$

The standard deviation of X is therefore $\sqrt{\frac{83}{324}} = 0.506$ (to 3 sf).

You should use fractions where possible and should only round the final answer.

S2

If the range of possible values of X is finite, then the range usually has a magnitude of between 3σ and 6σ.

Range is the difference between the largest and smallest possible values of X.

Example 14

Verify that the calculations in Example 13 seem reasonable.

The extremes of the data in Example 13 were 1 and 3. The value calculated for $E(X)$ was approximately 2.7. Since 2.7 lies between 1 and 3, the calculations seem reasonable.

The range of the possible values of X is $3 - 1 = 2$. This is about 4 times the calculated standard deviation (0.506), so again there is no suggestion of an incorrect calculation.

Exercise 1D

Find the probability distribution of X for each of the following cases. Hence, find the mean, variance and standard deviation of X.

1 A fair coin has the number '1' on one face and the number '2' on the other. The coin is thrown with a fair six-faced die and X is the sum of the scores.

2 Two fair dice, one red and the other green, are thrown and X is the score on the red die minus the score on the green die.

3 Two fair dice, one red and the other green, are thrown and X is the difference (possibly zero) between the larger and the smaller of the two scores.

4 A box contains three red marbles and five green marbles. Two marbles are taken at random *without* replacement, and X is the number of green marbles obtained.

5 A box contains three red marbles and five green marbles. Two marbles are taken at random *with* replacement, and X is the number of green marbles obtained.

6 Packets of 'Hidden Gold' cornflakes are sold for £1.20 each. One in twenty of the packets contains a £1 coin. A shopper buys two packets and £X is the net cost of the two packets.

7 In a raffle, 20 tickets are sold and there are two prizes. One ticket number is drawn at random and the corresponding ticket earns a £10 prize. A second, different, ticket number is drawn at random, and the corresponding ticket earns a £3 prize. The prize earned by a particular one of the original 20 tickets is £X.

Expectation (mean) of a function of a random variable

The idea that $\mathrm{E}(X)$ is the long-term average value of the random variable X, and that $\mathrm{E}(X^2)$ is the long-term average value of X^2 applies to any function of X.

For a general function, $\mathrm{g}(X)$, the value of $\mathrm{E}(\mathrm{g}(X))$ is calculated using:

$$\mathrm{E}(\mathrm{g}(X)) = \sum \mathrm{g}(x)\mathrm{P}(X = x) \qquad (1.8)$$

where the summation is over all possible values of X.

An equivalent form is:

$$\mathrm{E}(\mathrm{g}(X)) = \sum_i \mathrm{g}(x_i)p_i$$

Example 15

Calculate the expectation of $\frac{1}{X}$, where X is the value obtained from rolling a fair six-faced die.

Each of the six possible values is equally likely, and therefore:

$$\mathrm{E}\left(\frac{1}{X}\right) = \left(\frac{1}{1} \times \frac{1}{6}\right) + \left(\frac{1}{2} \times \frac{1}{6}\right) + \cdots + \left(\frac{1}{6} \times \frac{1}{6}\right)$$
$$= \frac{1}{6} + \frac{1}{12} + \cdots + \frac{1}{36}$$
$$= \frac{147}{360} = \frac{49}{120}$$

S2

The expectation of $\frac{1}{X}$ is about 0.4 and is *not* simply the reciprocal of $\mathrm{E}(X)$ (which would have given the answer $\frac{2}{7}$).

In general, $\mathrm{E}(\mathrm{g}(X)) \neq \mathrm{g}(\mathrm{E}(X))$

Example 16

The discrete random variable X is equally likely to take the values 0, 1 or 2. Determine the expected value of $3X^2$.

$$\mathrm{E}(3X^2) = (3 \times 0^2 \times \mathrm{P}(X = 0)) + (3 \times 1^2 \times \mathrm{P}(X = 1)) + (3 \times 2^2 \times \mathrm{P}(X = 2))$$
$$= 0 + \left(3 \times 1^2 \times \frac{1}{3}\right) + \left(3 \times 2^2 \times \frac{1}{3}\right)$$
$$= 0 + 1 + 4 = 5$$

The expected value of $3X^2$ is 5.

Example 17

The random variable X is equally likely to take the values 1, 2, and 4, and takes no other values.

Determine the mean and variance of $16X^{-2}$.

It is helpful to write $16X^{-2} = \frac{16}{X^2} = Y$. Then,

when $X = 1$,

$$Y = \frac{16}{1^2} = \frac{16}{1} = 16$$

when $X = 2$,

$$Y = \frac{16}{2^2} = \frac{16}{4} = 4$$

and when $X = 4$,

$$Y = \frac{16}{4^2} = \frac{16}{16} = 1$$

Since X is equally likely to take the values 1, 2 and 4, it follows that Y (that is, $16X^{-2}$) is equally likely to take the values 16, 4 and 1.

The mean of Y is given by:

$$E(Y) = \left(16 \times \frac{1}{3}\right) + \left(4 \times \frac{1}{3}\right) + \left(1 \times \frac{1}{3}\right) = \frac{21}{3} = 7$$

Similarly, $E(Y^2)$ is given by:

$$E(Y^2) = \left(16^2 \times \frac{1}{3}\right) + \left(4^2 \times \frac{1}{3}\right) + \left(1^2 \times \frac{1}{3}\right)$$

$$= \frac{256 + 16 + 1}{3} = \frac{273}{3} = 91$$

Thus $16X^{-2}$ has mean 7 and variance given by

$$\text{Var}(Y) = E(Y^2) - (E(Y))^2 = 91 - 7^2 = 42$$

S2

Exercise 1E

1 The random variable X has distribution given by:

x	0	1	4
$P(X = x)$	$\frac{1}{4}$	$\frac{7}{12}$	$\frac{1}{6}$

Find:

a) the mean of X b) $E(X^2)$ c) $E(3X + 2)$.

2 The random variable Y has distribution given by:

y	0	1	4
$P(Y = y)$	$\frac{1}{4}$	$\frac{1}{4}$	$\frac{1}{2}$

a) Find:
i) $E(Y)$ ii) $E(Y^2)$ iii) $E(\sqrt{Y})$.

b) Verify that $E(Y) > (E(\sqrt{Y}))^2$.

3 The random variable X has distribution given by:

x	−1	3	8
$P(X = x)$	$\frac{1}{5}$	$\frac{2}{5}$	$\frac{2}{5}$

a) Find:
i) $E(X)$ ii) $E(X^2)$ iii) $E(X^{-1})$

b) Determine whether $E(X^{-1}) = (E(X))^{-1}$.

4 The random variable X has distribution given by:
$P(X = 1) = 0.4$, $P(X = 4) = 0.6$.

a) Determine the mean and variance of X.

b) Determine the mean and variance of Y, where $Y = \sqrt{X}$.

5 The random variable X has distribution given by:
$P(X = k) = 0.4$, $P(X = 1) = 0.6$.
The random variable Y is given by $Y = \frac{1}{2}X^2$.

a) Determine the mean and variance of Y for the case $k = 0$.

b) Determine the mean and variance of Y for the case $k = -1$.

S2

1.3 Expectation (mean), variance and standard deviation of simple functions of a random variable

This section gives some very useful results that are concerned with simple transformations of a random variable. Where appropriate, the simplifying notation $p_i = P(X = x_i)$ is used.

Example 18

Suppose that the discrete random variable X has probability distribution given by:

$$P(X = 0) = P(X = 1) = 0.4 \qquad P(X = 2) = 0.2$$

The random variable Y is defined by $Y = 2X - 1$.

a) Determine the mean and variance of X and of Y.

b) Comment on the results.

a) The simplest approach is to make a table of the probabilities and the possible values for X and Y:

p_i	0.4	0.4	0.2
x_i	0	1	2
$y_i = 2x_i - 1$	−1	1	3

Note that $P(X = x_i) = P(Y = y_i)$.

$$E(X) = (0 \times 0.4) + (1 \times 0.4) + (2 \times 0.2) = 0.8$$
$$E(Y) = ((-1) \times 0.4) + (1 \times 0.4) + (3 \times 0.2) = 0.6$$

In order to obtain the variances, values are required for $E(X^2)$ and $E(Y^2)$.

Calculate the variance of X using $Var(X) = E(X^2) - (E(X))^2$, (and similarly for the variance of Y).

$$E(X^2) = (0^2 \times 0.4) + (1^2 \times 0.4) + (2^2 \times 0.2) = 1.2$$
$$E(Y^2) = ((-1)^2 \times 0.4) + (1^2 \times 0.4) + (3^2 \times 0.2) = 2.6$$

Hence:

$$Var(X) = 1.2 - 0.8^2 = 1.2 - 0.64 = 0.56$$
$$Var(Y) = 2.6 - 0.6^2 = 2.6 - 0.36 = 2.24$$

b) Thus

$$E(Y) = 0.6$$
$$2E(X) - 1 = 2 \times 0.8 - 1 = 0.6$$

so:

$$E(Y) = E(2X - 1) = 2E(X) - 1$$

Similarly

$$Var(Y) = 2.24$$
$$2^2 \times Var(X) = 4 \times 0.56 = 2.24$$

showing that:

$$Var(Y) = Var(2X - 1) = 2^2 \times Var(X)$$

These connections between E(X) and E(Y) and between Var(X) and Var(Y) are examples of general results derived later.

S2

E($X + b$) and Var($X + b$)

Suppose that the random variable X refers to the distance (in cm) between the top of a person's head and ground level. A group of three people is illustrated below.

The three people now stand on a platform that is 20 cm high, as shown below.

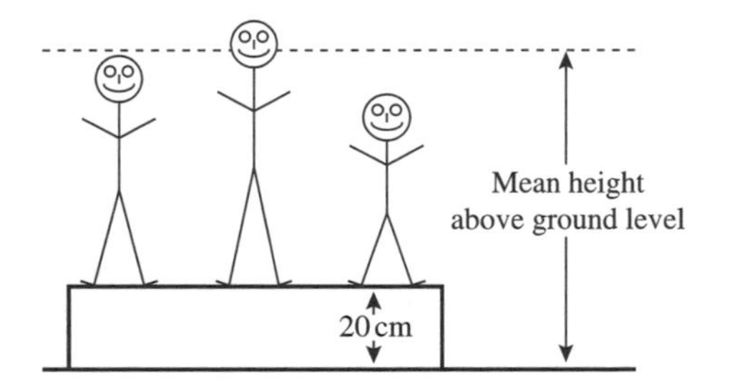

Since all are standing on the platform, the value of X for each person has increased by 20 cm and therefore the mean value of X has increased by 20 cm. However, the new values of X are no more variable than the old ones, since the individual differences from the mean height are the same as they were previously. Generalising these results to a platform of height b cm gives:

$$E(X + b) = E(X) + b \qquad (1.9)$$
$$Var(X + b) = Var(X) \qquad (1.10)$$

These results mirror those for linear scaling of data in Section 1.3 of *Statistics S1*.

Suppose instead that the people now stand in a pit that is 30 cm deep.

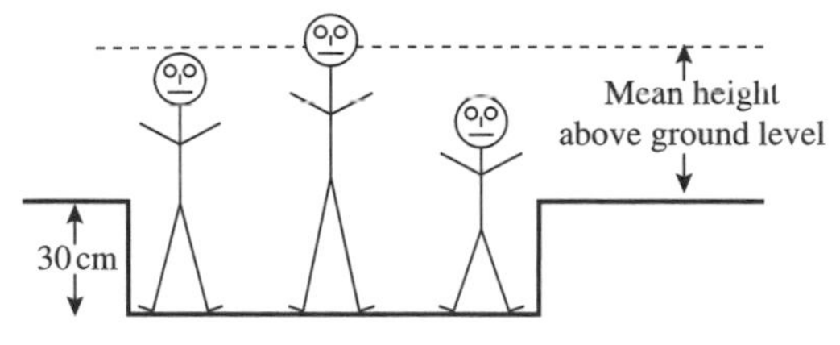

Each X value has been reduced by 30 cm, so the mean value of X is reduced by that amount. However, the variability of the values of X is again unaffected. Generalising to a pit of depth b cm gives:

$$\mathrm{E}(X - b) = \mathrm{E}(X) - b$$
$$\mathrm{Var}(X - b) = \mathrm{Var}(X)$$

These are really the same as (1.9) and (1.10), with b replaced by $-b$.

S2

These results are true for any random variable. Here is a proof. It is quite long, but is simpler than it appears at first sight. Let $Y = X + b$, where b is any constant (either positive or negative) and $\mathrm{P}(X = x_i)$ is denoted by p_i. Now:

$$\mathrm{E}(Y) = \mathrm{E}(X + b) = \sum(x_i + b)p_i$$

where the summation is over all possible values. Thus:

$$\begin{aligned} \mathrm{E}(X + b) &= \sum x_i p_i + \sum b p_i \\ &= \sum x_i p_i + b\sum p_i && \text{since } b \text{ is a constant} \\ &= \mathrm{E}(X) + b\sum p_i && \text{by definition of } \mathrm{E}(X) \\ &= \mathrm{E}(X) + b && \text{since } \sum p_i = 1 \end{aligned}$$

which proves the first result, (1.9).

Since:

$$\mathrm{Var}(Y) = \mathrm{E}(Y^2) - (\mathrm{E}(Y))^2$$

an expression is needed for $\mathrm{E}(Y^2)$. Substituting $(X + b)$ for Y, as before, gives:

$$\begin{aligned} \mathrm{E}(Y^2) &= \mathrm{E}((X + b)^2) \\ &= \mathrm{E}(X^2 + 2bX + b^2) \\ &= \sum(x_i^2 + 2bx_i + b^2)p_i \\ &= \sum x_i^2 p_i + \sum 2bx_i p_i + \sum b^2 p_i \\ &= \sum x_i^2 p_i + 2b\sum x_i p_i + b^2\sum p_i \\ &= \mathrm{E}(X^2) + 2b\mathrm{E}(X) + b^2 \end{aligned}$$

This uses the definitions of $\mathrm{E}(X)$ and $\mathrm{E}(X^2)$, the fact that b is constant, and the fact that $\sum p_i = 1$.

Finally:

$$\begin{aligned} \mathrm{Var}(Y) &= \mathrm{E}(Y^2) - (\mathrm{E}(Y))^2 \\ &= \mathrm{E}(X^2) + 2b\mathrm{E}(X) + b^2 - (\mathrm{E}(X) + b)^2 \\ &= \mathrm{E}(X^2) + 2b\mathrm{E}(X) + b^2 - ((\mathrm{E}(X))^2 + 2b\mathrm{E}(X) + b^2) \\ &= \mathrm{E}(X^2) + 2b\mathrm{E}(X) + b^2 - (\mathrm{E}(X))^2 - 2b\mathrm{E}(X) - b^2 \\ &= \mathrm{E}(X^2) - (\mathrm{E}(X))^2 \\ &= \mathrm{Var}(X) \end{aligned}$$

which proves the second result, (1.10).

$$E(X + b) = E(X) + b \qquad (1.11)$$

for any constant b (positive or negative).

You should learn these formulae.

$$\text{Var}(X + b) = \text{Var}(X) \qquad (1.12)$$

for any constant b (positive or negative).

Example 19

A lottery ticket costs 10p. There are 10 000 tickets for the lottery, which has a top prize of £100 and 10 runner-up prizes of £10 each. Determine the expected gain or loss resulting from the purchase of a ticket.

Let X be the random variable denoting the amount (in £) won by a ticket. Assuming all the tickets are sold, the probability distribution of X is given by:

x_i	100	10	0
p_i	0.0001	0.0010	0.9989

Hence:

$$E(X) = (100 \times 0.0001) + (10 \times 0.0010) + (0 \times 0.9989) = 0.02$$

Since the lottery ticket costs £0.10, the expected value of $Y = (X - 0.10)$ is required. Using the general result,

$$E(Y) = E(X - 0.10) = E(X) - 0.10 = -0.08$$

On average, therefore, the purchase of a ticket will result in a loss of 8p.

Of course this may not deter you! You can possibly afford to gamble 10p, even with these very unfavourable circumstances, for the slight chance of being £99.90 better off.

E(aX) and Var(aX)

For simplicity, suppose that $a = 2$ and that each of the people illustrated in the original diagram is one of a pair of identical twins, remarkably adept at gymnastics – as illustrated below.

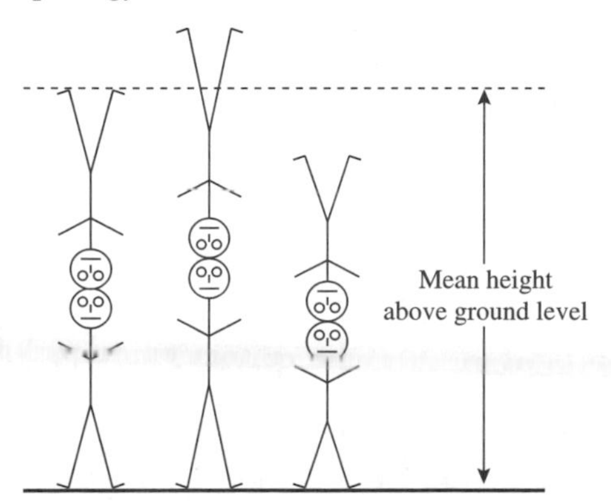

Let the random variable Y be the distance from the top twin's feet to ground level. Obviously, for each pair of twins, $Y = 2X$, where X is the height of one of the twins. So the mean of the Y-values must be double that of the X-values.

This is an example of linear scaling (discussed in Chapter 1 of *Statistics S1*).

The general result, true for any random variable and for any constant a (positive or negative) is easily obtained. Write $Y = aX$ and $y_i = ax_i$, so that:

$$\mathrm{P}(Y = y_i) = \mathrm{P}(aX = ax_i) = \mathrm{P}(X = x_i) = p_i$$

Then:

$$\mathrm{E}(Y) = \sum y_i p_i = \sum ax_i p_i = a\sum x_i p_i = a\mathrm{E}(X)$$

$$\mathrm{E}(aX) = a\mathrm{E}(X) \qquad (1.13)$$

for any constant a (positive or negative).

S2

It is obvious from the height diagrams that the Y-values are a great deal more variable than the X-values, but to find out exactly how much more variable requires some algebra.

From the definition of variance:

$$\mathrm{Var}(Y) = \mathrm{E}(Y^2) - (\mathrm{E}(Y))^2$$

Putting Y equal to aX, and using $\mathrm{E}(Y) = \mathrm{E}(aX) = a\mathrm{E}(X)$, gives:

$$\mathrm{Var}(Y) = \mathrm{E}((aX)^2) - (a\mathrm{E}(X))^2$$

But:

$$\mathrm{E}((aX)^2) = \sum (ax_i)^2 p_i = a^2 \sum x_i^2 p_i = a^2\mathrm{E}(X^2)$$

and

$$(a\mathrm{E}(X))^2 = a^2(\mathrm{E}(X))^2$$

Hence:

$$\mathrm{Var}(Y) = a^2\mathrm{E}(X^2) - a^2(\mathrm{E}(X))^2 = a^2(\mathrm{E}(X^2) - (\mathrm{E}(X))^2)$$
$$= a^2\mathrm{Var}(X)$$

and thus:

$$\mathrm{Var}(aX) = a^2\mathrm{Var}(X) \qquad (1.14)$$

for any constant a (positive or negative).

E($aX + b$) and Var($aX + b$)

Combining the previous results:

$$\mathrm{E}(aX + b) = \mathrm{E}(aX) + b = a\mathrm{E}(X) + b \qquad (1.15)$$
$$\mathrm{Var}(aX + b) = \mathrm{Var}(aX) = a^2\mathrm{Var}(X) \qquad (1.16)$$

for any constants a and b (positive or negative).

You should learn these formulae.

Example 20

At a fairground there is the following game. The player pays 20p in order to toss three coins. The stall-holder pays the player (in pence) 10 times the number of Heads that the player obtains. Determine a) the mean and b) the standard deviation of the player's net loss.

a) Let X be the random variable denoting the number of Heads obtained. The net loss (in pence) is given by $Y = 20 - 10X$.

A probability tree diagram for the outcomes of the three tosses is as follows:

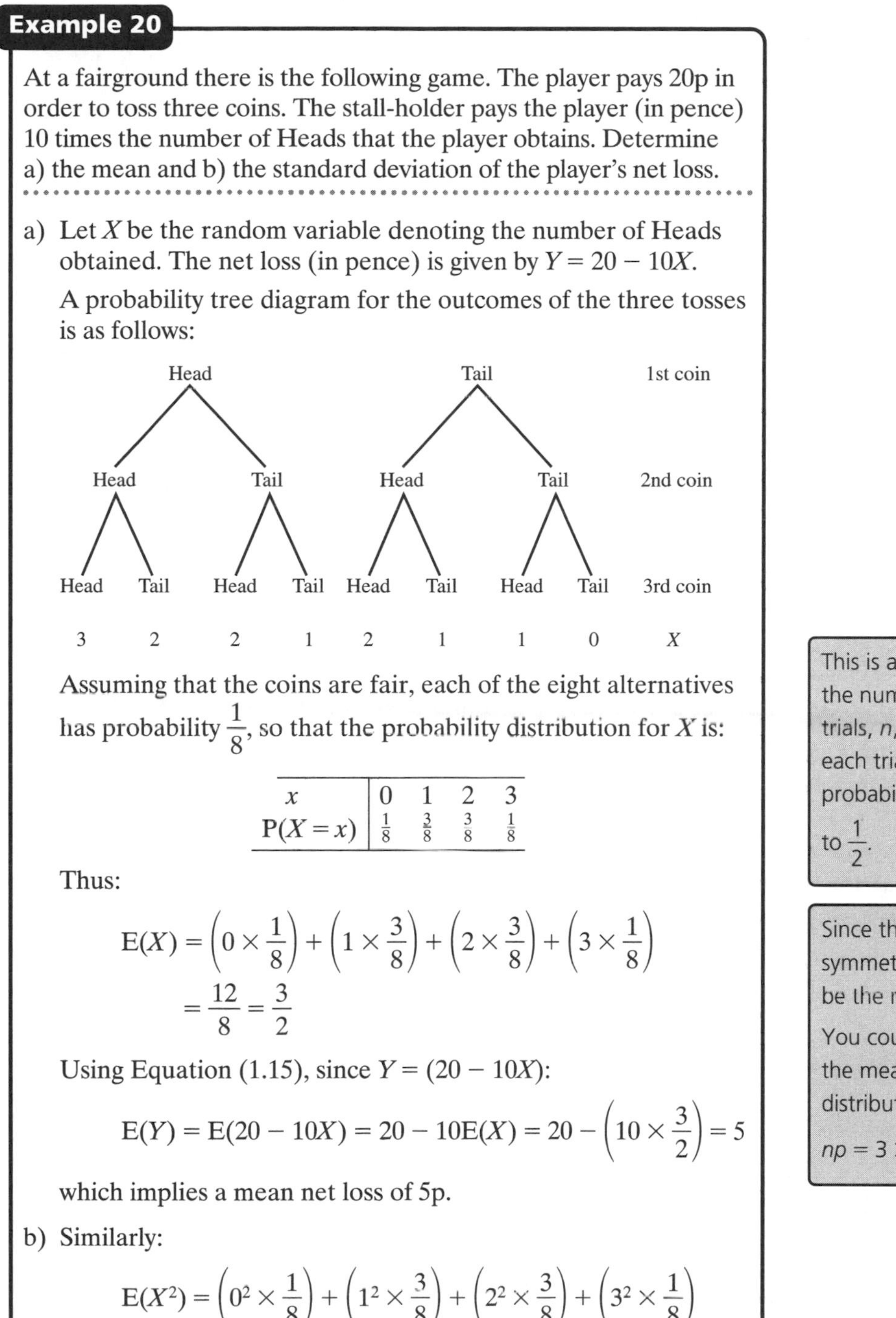

Assuming that the coins are fair, each of the eight alternatives has probability $\frac{1}{8}$, so that the probability distribution for X is:

x	0	1	2	3
$\mathrm{P}(X = x)$	$\frac{1}{8}$	$\frac{3}{8}$	$\frac{3}{8}$	$\frac{1}{8}$

Thus:

$$\mathrm{E}(X) = \left(0 \times \frac{1}{8}\right) + \left(1 \times \frac{3}{8}\right) + \left(2 \times \frac{3}{8}\right) + \left(3 \times \frac{1}{8}\right)$$
$$= \frac{12}{8} = \frac{3}{2}$$

Using Equation (1.15), since $Y = (20 - 10X)$:

$$\mathrm{E}(Y) = \mathrm{E}(20 - 10X) = 20 - 10\mathrm{E}(X) = 20 - \left(10 \times \frac{3}{2}\right) = 5$$

which implies a mean net loss of 5p.

b) Similarly:

$$\mathrm{E}(X^2) = \left(0^2 \times \frac{1}{8}\right) + \left(1^2 \times \frac{3}{8}\right) + \left(2^2 \times \frac{3}{8}\right) + \left(3^2 \times \frac{1}{8}\right)$$
$$= \frac{24}{8} = 3$$

Hence:

$$\mathrm{Var}(X) = 3 - \left(\frac{3}{2}\right)^2 = \frac{3}{4}$$

This is a binomial situation, with the number of independent trials, n, equal to 3, and with each trial having the same probability of success, p, equal to $\frac{1}{2}$.

Since the distribution is symmetrical about $\frac{3}{2}$, this must be the mean.

You could also use the fact that the mean of a binomial distribution is np. Here $np = 3 \times \frac{1}{2} = \frac{3}{2} = 1.5$.

The variance of a binomial distribution is $npq = np(1 - p)$. Here, therefore, $3 \times \frac{1}{2}\left(1 - \frac{1}{2}\right) = \frac{3}{4}$.

S2

Using the general result of Equation (1.16):

$$\text{Var}(Y) = \text{Var}(20 - 10X) = (-10)^2\,\text{Var}(X) = 100 \times \frac{3}{4} = 75$$

Hence, the standard deviation of the player's net loss is $\sqrt{75}$p, which (to 2 sf) is 8.7p. This correctly indicates that the outcome of this game is very uncertain!

Variance revisited

Equation (1.7) defined Var(X) by:

$$\text{Var}(X) = \text{E}(X^2) - (\text{E}(X))^2$$

An equivalent definition is provided by:

$$\text{Var}(X) = \text{E}((X - \mu)^2) \qquad (1.17)$$

where μ is the mean, E(X).

Equation (1.17) is how variance was originally defined historically, but Equation (1.7) is more useful in calculations.

It is now apparent that Var(X) is the expected value of a squared quantity and therefore cannot be negative.

It follows that, unless Var(X) = 0, $\text{E}(X^2) > (\text{E}(X))^2$.

Example 21

Demonstrate that:

$$\text{E}((X - \mu)^2) = \text{E}(X^2) - (\text{E}(X))^2$$

where $\mu = \text{E}(X)$.

This can be shown in many ways. Here is one:

$$\begin{aligned}\text{E}((X - \mu)^2) &= \sum(x_i - \mu)^2 p_i \\ &= \sum(x_i^2 - 2\mu x_i + \mu^2)p_i \\ &= \sum x_i^2 p_i - 2\mu\sum x_i p_i + \mu^2\sum p_i \\ &= \text{E}(X^2) - 2\mu\text{E}(X) + \mu^2\end{aligned}$$

This uses the definitions of $\text{E}(X^2)$ and E(X), together with $\sum p_i = 1$.

But $\mu = \text{E}(X)$, and so, substituting:

$$\begin{aligned}\text{E}((X - \mu)^2) &= \text{E}(X^2) - 2\text{E}(X) \times \text{E}(X) + (\text{E}(X))^2 \\ &= \text{E}(X^2) - (\text{E}(X))^2\end{aligned}$$

This gives the required result.

Exercise 1F

1 Given that E(X) = 0 and Var(X) = 1, and $Y = 3X - 4$, find E(Y) and Var(Y).

2 Given that $\text{E}(Y) = \frac{3}{4}$ and $\text{Var}(Y) = \frac{1}{4}$, find $\text{E}\left(\frac{1}{3}Y\right)$ and $\text{Var}\left(\frac{1}{3}Y\right)$.

3 Given that $E(X) = 4$, $Var(X) = 2$, find:

a) $E(3X + 6)$,

b) $Var(3X + 6)$,

c) $E(6 - 3X)$,

d) $Var(6 - 3X)$.

4 Given that $E(X) = 3$, $Var(X) = 4$, find the expectation and variance of:

a) $X - 2$,

b) $3X + 1$,

c) $2 - 3X$.

5 Given that $E(3Y + 2) = 8$ and $Var(4 - 2Y) = 12$, find the expected value and the variance of Y.

6 The random variable U has mean 10 and standard deviation 5.

The random variable V is defined by $V = \frac{1}{2}(U + 5)$.

Find the mean and standard deviation of V.

7 The random variable X has mean -2, and standard deviation 9. Find:

a) $E(X^2)$,

b) $E((X + 2)^2)$.

8 It costs £30 to hire a car for the day, and there is a mileage charge of 10p per mile. The distance travelled in a day has expectation 200 miles and standard deviation 20 miles.

Find the expectation and standard deviation of the daily cost.

9 The random variable Y takes the values $-1, 0, 1$, with probabilities $\frac{1}{4}, \frac{1}{2}, \frac{1}{4}$ respectively.

Find $E(10Y + 10)$ and $Var(10Y + 10)$.

10 Find $E(2S - 6)$ and $Var(2S - 6)$, where S is the score resulting from a single throw of an unbiased six-faced die.

11 The random variable T has mean 5 and variance 16.

Find two pairs of values for the constants c and d such that $E(cT + d) = 100$ and $Var(cT + d) = 144$.

12 Given that $E(X) = \mu$ and $Var(X) = \sigma^2$, find two pairs of values for the constants a and b such that $E(aX + b) = 0$ and $Var(aX + b) = 1$.

13 Given that $E(X) = \mu$, $Var(X) = \sigma^2$, and a is a constant, show that $E((X - a)^2) = (\mu - a)^2 + \sigma^2$.

Hence show that, as a varies, $E((X - a)^2)$ is least when $a = \mu$ and find the least value.

Summary

You should now be able to ...	Check out
1 Construct a simple probability distribution.	**1** Two fair six-faced dice are rolled. One is numbered as usual. The other has three faces numbered 0 and three faces numbered 1. The variable of interest is X, the sum of the scores on the two dice. Determine the probability distribution of X.
2 Determine the mean, variance and standard deviation of a variable having a known probability distribution.	**2** Determine the mean, variance and standard deviation of the random variable X having probability distribution: x: 0, 1, 2, 3; $P(X=x)$: 0.1, 0.4, 0.3, 0.2
3 Understand the term expectation.	**3** Determine the expectation of X, where: $P(X=2)=0.2$; $P(X=3)=0.7$; $P(X=4)=0.1$.
4 Calculate the variance and standard deviation of X, given $E(X)$ and $E(X^2)$.	**4** Calculate the variance and standard deviation of X given that $E(X^2)=169$ and $E(X)=5$.
5 Calculate the expectation of simple functions of X.	**5** For the probability distribution in question **3**, determine: a) $E(2X+3)$, b) $E(X^3)$, c) $E(X^2-2)$.
6 Calculate the variance of simple functions of X.	**6** For the probability distribution in question **3**, determine: a) $Var(X-1)$, b) $Var(2X+3)$.

S2

Revision exercise 1

1 The probability distribution of a random variable X is given in the table.

x	2	3	4	5
$P(X=x)$	0.2	0.3	0.4	0.1

Calculate:

a) the mean of X,

b) the variance of X. *(AQA, 2003)*

2 a) A cube has faces labelled 1, 1, 2, 2, 3 and 3. The cube is to be rolled once, and each face is equally likely to appear on top.

Write down the probability that the number showing will be 1.

b) Two cubes, both identical to the cube described in part a), are each to be rolled once.

Find the probability that:

i) both of the numbers showing will be 1,

ii) neither of the numbers showing will be 1.

c) The random variable X is the sum of the two numbers showing when the two cubes are rolled.

i) Copy and complete the following table.

x	2	3	4	5	6
$P(X = x)$		$\frac{2}{9}$	$\frac{3}{9}$	$\frac{2}{9}$	$\frac{1}{9}$

ii) Calculate the mean and variance of X. *(AQA, 2003)*

3 The probability distribution of a random variable X is given in the table below.

x	-2	-1	0	1	2
$P(X = x)$	0.1	0.1	0.2	0.3	0.3

Calculate:

a) the mean of X,

b) the variance of X. *(AQA, 2004)*

4 The probability distribution of a random variable X is given in the table below.

x	1	2	3	4
$P(X = x)$	$\frac{1}{6}$	$\frac{1}{6}$	$\frac{1}{6}$	$\frac{1}{2}$

Calculate the mean and variance of X. *(AQA, 2004)*

5 The discrete random variable R is such that $E(R) = 3$ and $Var(R) = 1$.

a) Determine the mean and variance of $5(R - 1)$.

The probability distribution of R is:

$$P(R = r) = \begin{cases} \dfrac{r}{10} & r = 1, 2, 3, 4 \\ 0 & \text{otherwise} \end{cases}$$

b) Calculate the mean and variance of $12R^{-1}$. *(AQA, 2002)*

6 The probability distribution for the number of vehicles, V, involved in each minor accident on a particular stretch of road can be modelled as follows.

v	1	2	3	4	5
$P(V = v)$	0.15	0.45	0.20	0.15	0.05

a) Show that $E(V) = 2.5$ and $Var(V) = 1.15$.

b) The total cost, £C, of removing all the damaged vehicles following a minor accident is given by

$$C = 30V + 25$$

Determine the mean and variance of C. *(AQA, 2003)*

S2

7 The probability distribution for the discrete random variable R is given in the table.

r	1	2	3	4	5
$P(R = r)$	0.1	0.2	0.4	0.2	0.1

a) Given that $E(R) = 3$ and $Var(R) = 1.2$, find the mean and variance of $5(2R - 1)$.

b) i) Write down the probability distribution for $\frac{60}{R}$.

ii) Show that $E\left(\frac{60}{R}\right) = 24.2$.

iii) Given that $E\left(\frac{3600}{R^2}\right) = 759.4$, determine the variance of $\frac{60}{R}$. *(AQA, 2003)*

8 The probability distribution for the number, R, of unwrapped sweets in a tin is given in the table.

r	0	1	2	3	4
$P(R = r)$	0.1	0.2	0.4	0.2	0.1

a) Show that:

i) $E(R) = 2$,

ii) $Var(R) = 1.2$.

b) The number, P, of partially wrapped sweets in a tin is given by

$$P = 3R + 4.$$

Find values for $E(P)$ and $Var(P)$.

c) The total number of sweets in a tin is 200. Sweets are either correctly wrapped, partially wrapped or unwrapped.

i) Express C, the number of correctly wrapped sweets in a tin, in terms of R.

ii) Hence find the mean and variance of C. *(AQA, 2004)*

9 Doctor Patel works at a health centre. On every Monday morning she has appointments with 5 patients.

a) Experience suggests that on each Monday the number of these patients, who had not booked an appointment with her during the previous year, may be modelled by the random variable X. The probability distribution of X is given in the table.

x	$P(X = x)$
0	0.20
1	0.34
2	0.31
3	0.10
4	0.04
5	0.01

Find:

i) the mean of X,

ii) $E(X^2)$,

iii) the standard deviation of X.

b) Of all patients registered with Doctor Patel, 40% have not booked an appointment with her during the previous year. Five patients registered with Doctor Patel were selected at random.

i) Name the distribution which could provide a model for the number of patients, Y, in the sample, who had not booked an appointment with her during the previous year.

ii) Write down the mean and standard deviation of Y.

c) Give a reason why patients in part a), who had booked an appointment on a Monday morning, are unlikely to constitute a random sample of **all** Dr. Patel's patients, based on:

i) your answers to a) and b),

ii) the context described. *(AQA, 2003)*

2 The Poisson distribution

This chapter will show you how to

- ✦ Recognize situations for which the Poisson distribution is appropriate
- ✦ Calculate Poisson probabilities from the formula
- ✦ Calculate Poisson probabilities from a table
- ✦ Solve problems involving sums of independent Poisson random variables

This distribution is named after Siméon Denis Poisson (1781–1840), a French mathematician.

Before you start

S2

You should know how to ...	Check in
1 Work with factorials.	**1** Calculate: a) $3!$, b) $\frac{6!}{5!}$, c) $\frac{3 \times 2 \times 1}{3!}$.
2 Apply a recurrence relation.	**2** It is given that $x_1 = 2$, and that $x_i = 2x_{i-1}$ for $i = 2, 3, 4$. Find the values of x_2, x_3 and x_4.
3 Evaluate e^{-x} from a calculator.	**3** Evaluate, correct to 3 significant figures: a) e^{-1}, b) $e^{-2.3}$, c) $e^{-4.25}$.

2.1 Conditions for application of a Poisson distribution

The diagram shows the locations of 65 pine saplings growing in a square region of side 5.7 m.

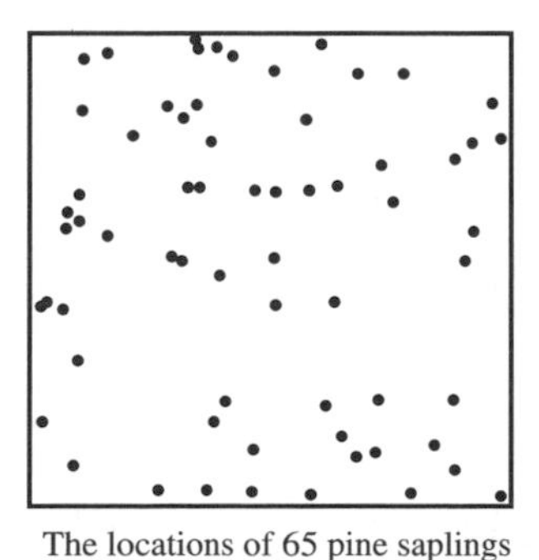

The locations of 65 pine saplings

The arrangement of the trees is typical of what is seen in nature. The trees were not deliberately planted in this way and nature's pattern appears entirely haphazard. Looking at the diagram, you cannot deduce where the saplings in the neighbouring plots will be found. This is a real-life example of data displaying an approximate Poisson process.

In a Poisson distribution the random variable is a count of events occurring *at random* in regions of time or space. 'At random' here has a very particular and strict definition: the occurrences of the events are required to be distributed through time or space so as to satisfy the following:

- ✦ Whether or not an event occurs at a particular point in time or space is independent of what happens elsewhere.
- ✦ At all points in time the probability of an event occurring within a small fixed interval of time is the same. This also applies to the probability of occurrence of events in small fixed-size regions of space.
- ✦ There is no chance of two events occurring at precisely the same point in time or space.

Any situation in which events obey these requirements is said to be a **Poisson process**. Typically, in a spatial Poisson process, there appear to be haphazardly arranged clusters of points as well as wide-open spaces.

'Spatial' refers to occurrences in space; 'temporal' refers to occurrences in time.

Examples of real-life Poisson processes are the following:

- ✦ The points in time at which a given piece of radioactive substance emits a charged particle.
- ✦ The points in space occupied by the micro-organisms in a random sample of well-stirred water taken from a pond.

Two somewhat bloodthirsty classic examples of observations from a Poisson distribution that appear prominently in older Statistics books are:

1. The numbers of deaths of cavalrymen caused by horsekicks! These data were collected with military precision each year for each of the various Prussian army corps during a period in the middle of the 19th century.
2. The numbers of bomb craters in equal-sized areas of wartime London.

A **Poisson distribution** describes the probabilities of the associated counts:

- ✦ The number of particles emitted in a minute by the radioactive substance.
- ✦ The number of micro-organisms in 1 ml of pond water.

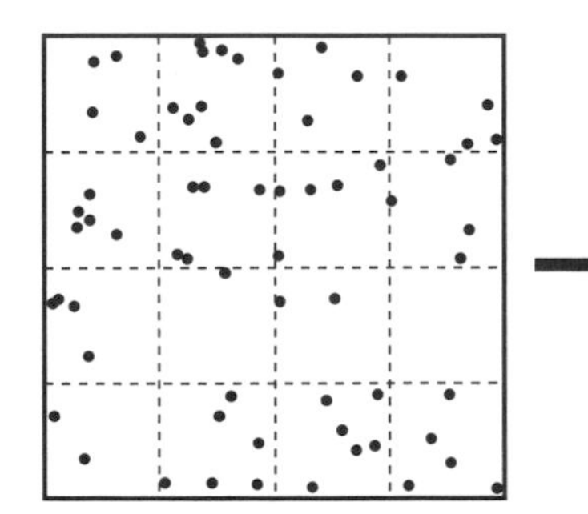

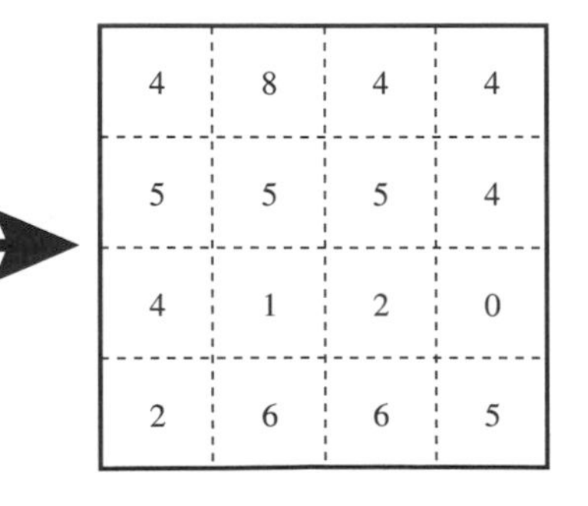

The way in which a Poisson distribution relates to a Poisson process is shown in the diagram. The figure shows a grid laid over the map of pine sapling locations (introduced on p 31) and the resulting counts of the numbers of trees in each grid square. If the trees are truly randomly located, then these will be observations from a Poisson distribution.

Example 1

The diagram shows the locations of sea anemones on a square region of rock at the side of a rock pool.

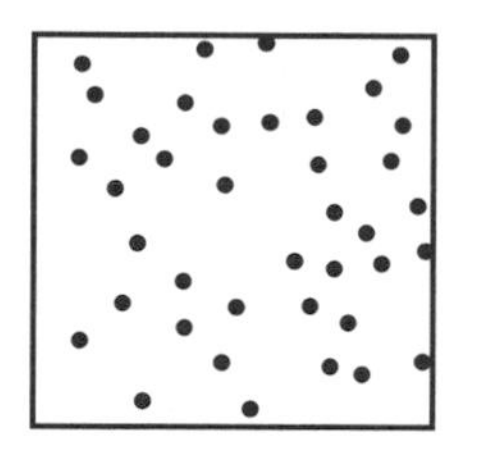

State, with a reason, whether it appears that the locations of these anemones are consistent with a Poisson process.

Obviously the anemones do not overlap! However, there is plenty of room on this rock, so each anemone can be thought of as being a point. There is no obvious pattern to the positions of anemones – they are not arranged in lines, nor are they packed together in a small part of the rock. It therefore seems reasonable to suppose that they are approximately randomly positioned: this is an example of a naturally occurring (approximate) Poisson process.

S2

Often, the locations are *deliberately* not Poisson. The following example illustrates this.

Example 2

The diagrams show a rough outline of Norwich and its immediate surroundings with points indicating the locations of schools and churches.

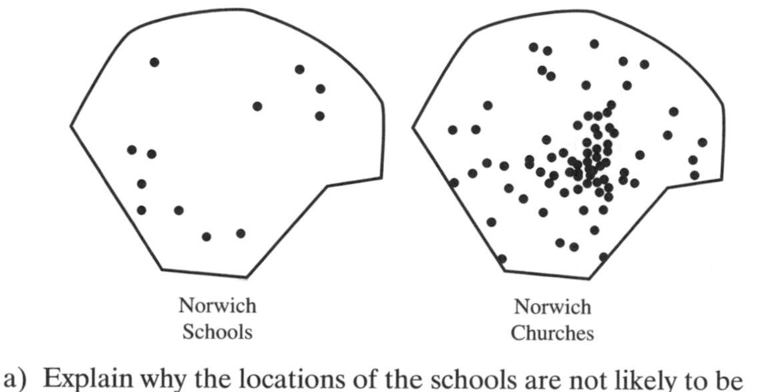

Norwich Schools

Norwich Churches

a) Explain why the locations of the schools are not likely to be consistent with a Poisson process.

b) Explain why the locations of the churches are not likely to be consistent with a Poisson process.

a) The Norwich schools are unlikely to have been placed randomly since they will have been rather regularly placed through the city suburbs so that no children live too far from school. When a new school is built the locations of existing schools are taken into account, so the independence property of the Poisson process is violated.

b) Most churches are old and are therefore more densely packed near the old-established centre of the city than in the more modern outskirts. This variation in the rate of occurrence contradicts the constant rate assumption of the Poisson process.

Exercise 2A

State, with a reason in each case, which of the following **might** give rise to observations from a Poisson distribution, which are **unlikely** to give rise to observations from a Poisson distribution, and which certainly **will not** give rise to observations from a Poisson distribution.

1 The numbers of phone calls received by a switchboard on randomly chosen days during June.

2 On a randomly chosen June day, the numbers of phone calls received by a switchboard during consecutive 5-minute periods from midnight until midnight the next day.

3 The numbers of cars passing in randomly chosen five-minute periods on a road with no traffic lights or long queues (assuming such a road exists!).

4 The numbers of cars passing in randomly chosen five-minute periods on a busy road in the city centre.

5 The numbers of currants in randomly chosen currant buns.

6 The numbers of currant buns in randomly chosen packs of six currant buns.

7 (There has been trouble in the currant-bun store: a mouse has been nibbling buns.) The numbers of currant buns, in randomly chosen packs of six currant buns, that have been nibbled by the mouse.

8 The numbers of accidents in a factory during a randomly chosen week.

9 The numbers of daisies in randomly chosen square metres of playing field.

10 The numbers of lions in randomly chosen one hectare regions of the African bush.

11 The numbers of faults in each of the cloth remnants in a curtain store.

12 The numbers of faults in randomly chosen square metres of cloth.

13 The numbers of typing errors on randomly chosen pages of the manuscript of a novel.

14 The numbers of errors on randomly chosen pages of a newspaper.

2.2 Calculation of probabilities using formula

The Poisson formula

The formula for a Poisson distribution involves one of the 'magic numbers' of mathematics, the number e = 2.718 28 … :

$$p_x = \mathrm{P}(X = x) = \begin{cases} e^{-\lambda}\dfrac{\lambda^x}{x!} & x = 0, 1, 2, \ldots \\ 0 & \text{otherwise} \end{cases} \qquad (2.1)$$

where λ, a Greek letter pronounced 'lambda', is a positive number. The corresponding shorthand is: $X \sim \mathrm{Po}(\lambda)$.

The number 'e', known as the exponential constant, is discussed in detail in the core mathematics module C3.

S2

$p_0 = e^{-\lambda}$ (since $0! = 1$, and $\lambda^0 = 1$ for all values of $\lambda > 0$).

The number λ is often referred to as the **parameter** of the distribution. This parameter represents the *rate* at which events occur (on average) in the region of time or space under consideration. Thus λ is always positive.

You should be careful to choose the correct value for λ. For example, if events occur at the rate 3 per hour, and the period of interest is 2 hours, then the value for λ will be $3 \times 2 = 6$ and not 3.

One way of defining the value of e is via the expression:

$$e^x = 1 + \frac{x}{1!} + \frac{x^2}{2!} + \frac{x^3}{3!} + \frac{x^4}{4!} + \cdots$$

so that:

$$e = e^1 = 1 + \frac{1}{1} + \frac{1}{2} + \frac{1}{6} + \frac{1}{24} + \cdots$$

You can use the definition of e^x, with x replaced by λ, to verify that the probabilities of the Poisson distribution do indeed sum to 1:

$$p_0 + p_1 + p_2 + \cdots = e^{-\lambda}\left(1 + \frac{\lambda}{1!} + \frac{\lambda^2}{2!} + \cdots\right) = e^{-\lambda}e^{\lambda} = 1$$

Exercise 2B

1 Using a calculator, determine the following to three significant figures:

a) $e^{-0.3}$, b) $e^{-1.4}$, c) $e^{-1.5}$, d) e^{-2}.

2 Using a calculator, determine the following to three significant figures:

a) $2e^{-1.5}$, b) $3e^{-0.3} + 0.4$.

Direct calculation of Poisson probabilities

You can see how to calculate Poisson probabilities by studying the next two examples.

Example 3

Between 6 pm and 7 pm, a directory enquiries firm receives calls at an average rate of 2 per minute.
The calls may be assumed to arrive at random points in time.

a) Suggest an appropriate distribution for the number of calls in a randomly chosen period of time between 6 pm and 7 pm, giving a reason for your answer.

b) Determine the probability that 4 calls arrive in a randomly chosen minute.

c) Determine the probability that fewer than 3 calls arrive in a randomly chosen two-minute period between 6 pm and 7 pm.

S2

Watch out: 'fewer than 3' means '2 or less'.

a) Since calls arrive at random points in time, a Poisson process is being described.

b) Let X be the number of calls that arrive in a randomly chosen minute.
Since the average rate at which calls arrive is 2 per one-minute period, $\lambda = 2$. Hence:

$$P(X = 4) = e^{-2}\frac{2^4}{4!} = 0.090 \text{ (to 3 dp)}$$

The probability that 4 calls arrive in a randomly chosen minute is 0.090 (to 3dp).

c) The average rate is 2 calls per minute, which implies that the average rate per two minutes is 4.
Thus, with Y denoting the number of calls that arrive in a randomly chosen two-minute period, $Y \sim \text{Po}(4)$ and hence:

$$\begin{aligned}P(Y < 3) &= P(Y = 0) + P(Y = 1) + P(Y = 2)\\ &= e^{-4}\frac{4^0}{0!} + e^{-4}\frac{4^1}{1!} + e^{-4}\frac{4^2}{2!}\\ &= e^{-4} \times \left(\frac{1}{1} + \frac{4}{1} + \frac{16}{2}\right)\\ &= 13e^{-4}\\ &= 0.238 \text{ (to 3 dp)}\end{aligned}$$

The probability that fewer than 3 calls arrive in a randomly chosen two-minute period is 0.238 (to 3 dp).

Example 4

In a certain disease a small proportion of the red blood corpuscles display a tell-tale characteristic. A test consists of taking a random sample of 2 ml of a person's blood and counting the number of distinctive corpuscles. A count of five or more is taken to be an indication that the person has the disease. Mrs Wretched has the disease; the distinctive corpuscles occur in her blood at an average rate of 1.6 per ml.

a) State a necessary assumption for the number of distinctive corpuscles in a random sample of her blood to have a Poisson distribution.

b) Determine the probability that a random sample of 2 ml of her blood will contain five or more of the distinctive corpuscles.

c) Does the test appear to be a good one?

a) A necessary assumption is that the corpuscles are randomly distributed through Mrs Wretched's blood.

b) Let X be the number of distinctive corpuscles in a random sample of 2 ml of her blood. Since the average rate of occurrence is 3.2 per 2 ml, the Poisson distribution has $\lambda = 3.2$. In order to calculate $P(X \geqslant 5)$ it is necessary to calculate the probability of the complementary event, $(X \leqslant 4)$:

$$P(X \leqslant 4) = p_0 + \cdots + p_4$$
$$= e^{-3.2}\left(1 + \frac{(3.2)^1}{1!} + \frac{(3.2)^2}{2!} + \frac{(3.2)^3}{3!} + \frac{(3.2)^4}{4!}\right)$$
$$= 0.781 \text{ (to 3 dp)}$$

Hence: $P(X \geqslant 5) = 1 - 0.781 = 0.219$

Thus the probability that 2 ml of Mrs. Wretched's blood contains five or more of the distinctive corpuscles is 0.219 (to 3 dp).

c) Since there is a chance of nearly 80% that the test will fail to suggest that Mrs. Wretched has the disease, the test is not very good.

The complementary event is used so as to avoid having to calculate, and add together, an infinite number of probabilities!

S2

Exercise 2C

In Questions 1–5, the random variable X has a Poisson distribution with parameter λ.

1 Given that $\lambda = 2$, find:

a) $P(X = 0)$, b) $P(X = 1)$, c) $P(X = 2)$,
d) $P(X \leqslant 2)$, e) $P(X \geqslant 2)$.

2 Given that $\lambda = 0.5$, find:

a) $P(X < 3)$, b) $P(2 \leqslant X \leqslant 4)$, c) $P(X \geqslant 3)$.

3 Given that $\lambda = 5$, find:

a) $P(X = 5)$, b) $P(X < 5)$, c) $P(X > 5)$.

4 Given that $\lambda = 1.4$, find $P(X = 1, 3 \text{ or } 5)$.

5 Given that $\lambda = 2.2$ and $P(X = x) = 0.1967$, find the value of x.

6 The number of currants in a randomly chosen currant bun can be modelled as a random variable having a Poisson distribution with parameter 5.6.

Find the probability that a randomly chosen currant bun contains:

a) less than three currants, b) three currants,

c) more than three currants, d) at least three currants.

7 The number of accidents in a randomly chosen week at a factory can be modelled by a Poisson distribution with parameter 0.7.

Find the probability that there are more than two accidents in a randomly chosen week.

8 The number of emergency calls received by a Gas Board in a randomly chosen day can be modelled by a Poisson distribution with parameter 3.4.

Find the probability that, in a randomly chosen day, the number of emergency calls received is 5 or more.

9 Buttercups are randomly distributed across a playing field with the probability that a randomly chosen square metre of the field contains precisely r buttercups being

$$e^{-2}\frac{2^r}{r!}, \qquad r = 0, 1, 2, \ldots$$

Determine the probability that a randomly chosen region of area 0.5 square metres contains precisely one buttercup.

Calculation of probabilities using a recurrence formula

If a question requires calculation of the probabilities for a succession of values of x, then there is a simple way of obtaining $P(X = x)$ from $P(X = x - 1)$.

Since, for $x > 1$:

$$p_x = \frac{\lambda^x}{x!}e^{-\lambda} = \frac{\lambda(\lambda^{x-1})}{x(x-1)!}e^{-\lambda} = \frac{\lambda}{x} \times \frac{\lambda^{x-1}}{(x-1)!}e^{-\lambda}$$

and

$$p_{x-1} = \frac{\lambda^{x-1}}{(x-1)!}e^{-\lambda}$$

it follows that:

$$p_x = \frac{\lambda}{x}p_{x-1}$$

So:

$$P(X=x) = \frac{\lambda}{x}P(X=x-1)$$

This gives a convenient **recurrence formula** for calculating a set of Poisson probabilities relatively quickly by starting with $P(X=0) = p_0 = e^{-\lambda}$.

Example 5

The random variable X has a Poisson distribution with parameter 1.7. Determine $P(X>3)$.

This question is answered by first calculating the probability of the complementary event, $X \leqslant 2$. Starting with p_0 and using the recurrence formula gives:

$$p_0 = e^{-1.7}(= 0.182\,683\,5)$$

$$p_1 = \frac{1.7}{1}p_0(= 0.310\,562\,0)$$

$$p_2 = \frac{1.7}{2}p_1(= 0.263\,977\,7)$$

$$p_3 = \frac{1.7}{3}p_2(= 0.149\,587\,4)$$

$$\begin{aligned} P(X>3) &= 1-(p_0+p_1+p_2+p_3) \\ &= 1-0.906\,810\,6 \\ &= 0.093 \text{ (to 3 dp)} \end{aligned}$$

Avoid rounding during the intermediate stages, otherwise the final answer is likely to be inaccurate.

Try it and see for yourself.

S2

Calculator practice

Some calculators can calculate individual, as well as cumulative, Poisson probabilities. The details will be given in the manual for the calculator. Typically you will need to switch to a special mode and select from the options available. You will then need to specify the value of λ, as well as the set of values of x for which the probability is required.

Exercise 2D

In Questions 1–4 the random variable X has a Poisson distribution with parameter λ. You should give your answers to three decimal places.

1 Given that $\lambda = 2$, calculate $P(X=0)$ and use the recurrence relation to calculate $P(X=x)$ for $x = 1, 2, 3, 4$.

Check by calculating $P(X=4)$ directly.

2 Given that $\lambda = 5$, calculate $P(X=4)$ directly and use the recurrence relation to calculate $P(X=x)$ for $x = 5, 6, 7$.

3 Given that $\lambda = 2.5$, calculate $P(X = 3)$ directly and use the recurrence relation (backwards) to calculate $P(X = x)$ for $x = 2$, 1, 0.

Check by calculating $P(X = 0)$ directly.

4 Suppose that, for a positive integer x,

$$P(X = x - 1) \leqslant P(X = x) \text{ and } P(X = x + 1) < P(X = x).$$

a) Determine, in terms of x, the possible range of values for λ.

b) Hence, state the value of x when:
i) $\lambda = 3.6$, ii) $\lambda = 5$.

2.3 Calculation of probabilities using tables

S2

Table 2 of the AQA statistical tables gives the values of the **cumulative** Poisson distribution function, $P(X \leqslant x)$, for various values of λ. Here is an extract:

λ	1.0	1.2	1.4	1.6	1.8	λ
x						x
0	0.3679	0.3012	0.2466	0.2019	0.1653	0
1	0.7358	0.6626	0.5918	0.5249	0.4628	1
2	0.9197	0.8795	0.8335	0.7834	0.7306	2
3	0.9810	0.9662	0.9463	0.9212	0.8913	3
4	0.9963	0.9923	0.9857	0.9763	0.9636	4
5	0.9994	0.9985	0.9968	0.9940	0.9896	5
6	0.9999	0.9997	0.9994	0.9987	0.9974	6
7	1.0000	1.0000	0.9999	0.9997	0.9994	7
8			1.0000	1.0000	0.9999	8
9					1.0000	9

You will be provided with these tables in the examination, and Table 2 is reproduced as Table 2 of the Appendices to this book.

Entries given as l.0000, and all gaps in the table, correspond to probabilities exceeding 0.999 95.

This extract shows $P(X \leqslant x)$, to four decimal places, for $x = 0, 1, \ldots, 9$ and for selected values of λ between 1.0 and 1.8, inclusive. The complete table gives cumulative probabilities for selected values of λ between 0.1 and 15.0, inclusive.

If a probability is required for a value of λ that is not included in the tables then (if it is not a special function on your calculator) it should be calculated from the formula.

The table is easy to use. For example, with $\lambda = 1.4$, you can see that $P(X \leqslant 4) = 0.9857$.

In order to use the table to find probabilities for individual values of r, or for other types of inequality, the following relations are required:

Relation	Example
$P(X < x) = P(X \leqslant x - 1)$	$P(X < 6) = P(X \leqslant 5)$
$p_x = P(X = x) = P(X \leqslant x) - P(X \leqslant x - 1)$	$P(X = 4) = P(X \leqslant 4) - P(X \leqslant 3)$
$P(X > x) = 1 - P(X \leqslant x)$	$P(X > 2) = 1 - P(X \leqslant 2)$
$P(X \geqslant x) = 1 - P(X \leqslant x - 1)$	$P(X \geqslant 5) = 1 - P(X \leqslant 4)$

Be careful not to confuse $P(X < x)$ with $P(X \leqslant x)$.

You should be familiar with these relations, since they were also required for the binomial distribution (Statistics S1, Chapter 3).

Example 6

Tadpoles are scattered randomly through a pond at the average rate of 14 a litre. A random sample of 0.1 litres is examined.

a) State, with a reason, the type of distribution that is appropriate.

b) State the average number of tadpoles for a random sample of 0.1 litres.

c) Determine the probability that the sample will contain more than 3 tadpoles.

a) Since the tadpoles are distributed at random through the pond, a Poisson distribution is appropriate.

b) Since the average rate is 14 per litre, the average number in 0.1 litres is $0.1 \times 14 = 1.4$.

c) The probability that there are more than 3 in the sample is obtained by using:

$$P(X > 3) = 1 - P(X \leqslant 3)$$

From Table 2, with $\lambda = 1.4$, $P(X \leqslant 3) = 0.9463$. Thus the required probability is:

$$\begin{aligned} P(X > 3) &= 1 - 0.9463 \\ &= 0.0537 \end{aligned}$$

The probability that the sample will contain more than 3 tadpoles is 0.054 (to 3 dp).

Example 7

The random variable Y has a Poisson distribution with parameter 1.8. Determine the probability that Y takes a value greater than 4, but less than 7.

The question is asking for:

$$p_5 + p_6$$

Using Table 2, for $\lambda = 1.8$:

$$\begin{aligned} P(Y \leqslant 6) - P(Y \leqslant 4) &= 0.9974 - 0.9636 \\ &= 0.0338 \end{aligned}$$

and so the probability that Y takes a value greater than 4, but less than 7, is 0.034 (to 3 dp).

Example 8

Use tables of the cumulative Poisson distribution function to determine $P(1 < X \leqslant 5)$, where X is a Poisson random variable with parameter 1.0.

The question requires:

$$p_2 + p_3 + p_4 + p_5$$

Using Table 2, with $\lambda = 1.0$:

$$P(X \leqslant 5) - P(X \leqslant 1) = 0.9994 - 0.7358 = 0.2636$$

The required probability is 0.264 (to 3 dp).

A random variable with a Poisson distribution is often referred to as a Poisson random variable.

S2

Exercise 2E

In Questions 1–6 the random variable X has a Poisson distribution with parameter λ. Use Table 2 to answer the following questions.

1 Given that $\lambda = 3$, find:

a) $P(X \leqslant 5)$, b) $P(X < 7)$.

2 Given that $\lambda = 0.9$, find:

a) $P(X \geqslant 3)$, b) $P(X > 4)$.

3 Given that $\lambda = 1.2$, find:

a) $P(2 < X < 5)$, b) $P(2 \leqslant X \leqslant 5)$.

4 Given that $\lambda = 1$, find:

a) $P(X = 1)$, b) $P(X = 2)$.

5 Find λ, given that $P(X \leqslant 5) = 0.9896$.

6 Find λ, given that $P(X > 4) = 0.0527$.

7 The number of emergency calls received by a town's fire service in any week can be modelled by a Poisson distribution with mean 15.

Calculate the probability that, in any week, the number of emergency calls received by this fire service is exactly 20.

8 The number of people joining a checkout queue at a supermarket may be modelled by a Poisson distribution with a mean of 1.8 per minute. Find the probability that in a particular minute the number of people joining the queue is:

a) one or fewer, b) exactly three. *(AQA, 2003)*

9 The weekly number of ladders sold by a small DIY shop can be modelled by a Poisson distribution with mean 1.4. Find the probability that in a particular week the shop will sell:

a) 2 or fewer ladders, b) exactly 4 ladders,

c) 2 or more ladders. *(AQA, 2002)*

10 Messages by email are received by Tariq's personal computer, independently, at random, at an average rate of 2.4 per day.

a) Name a distribution which provides a suitable model for the number of email messages received per day by Tariq's personal computer.

b) Find the probability that the number of messages received on a particular day is:

i) two or fewer, ii) exactly four.

c) Find the probability that the number of messages received during a five-day period is:

i) fewer than six, ii) more than 17.

(*AQA, 2004*)

2.4 Mean, variance and standard deviation of a Poisson distribution

Since the parameter λ is the average rate of occurrence of events in the amounts of space or time that interest us, it is also the mean of the distribution. However, for the Poisson distribution, λ is not only the mean, it is also the variance. Thus $\sqrt{\lambda}$ is the standard deviation.

The proof of these results is not part of A2 Module Statistics 2 and is therefore omitted here.

A Poisson distribution with parameter λ has mean λ, variance λ, and standard deviation $\sqrt{\lambda}$.

Since λ refers to a rate of occurrence it is always positive.

The shape of a Poisson distribution

The bar charts show the probability of values of X where $X \sim \text{Po}(\lambda)$, for varying λ.

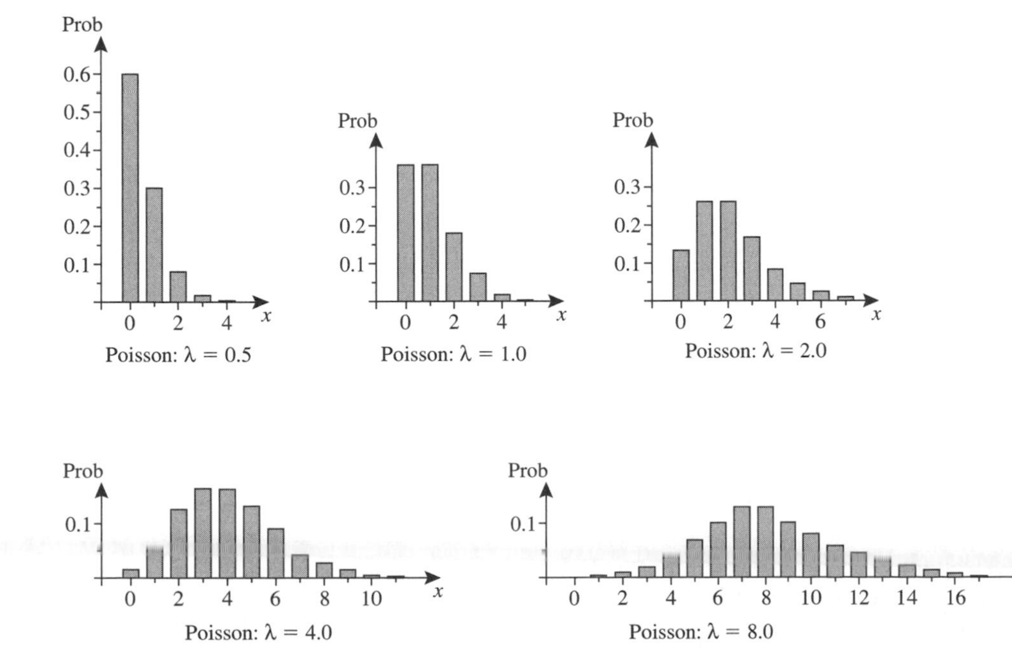

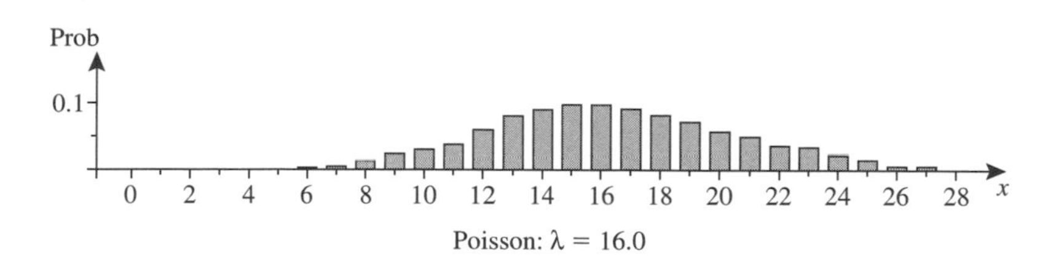

Poisson: $\lambda = 16.0$

When λ is an integer the distribution has modes at $x = \lambda$ and $x = (\lambda - 1)$. Otherwise the mode occurs when x is equal to the largest integer less than λ. Thus, for $\lambda < 1$ the distribution has mode at $x = 0$ and is very skewed. As λ increases so the distribution takes on a more symmetrical appearance.

> Remember that there is no upper bound on the value of a Poisson random variable. The bar charts stop when the bars become difficult to see!

S2

Although the range of possible values is infinite, in practice (for large λ) about 95% of values will lie between $\lambda - 2\sqrt{\lambda}$ and $\lambda + 2\sqrt{\lambda}$ (that is, in the range: mean $\pm$ 2 standard deviations).

2.5 Distribution of a sum of independent Poisson random variables

> If X and Y are independent Poisson random variables with parameters λ and ϕ, respectively, then the random variable S, defined by $S = X + Y$ is a Poisson random variable with parameter $\lambda + \phi$.

A direct proof of this result is tedious, but the result is almost obvious once you consider the Poisson process background to the distribution.

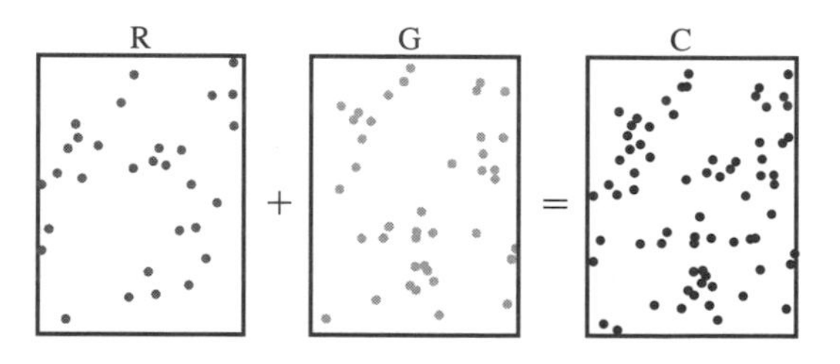

A Poisson distribution refers to counts of 'events' scattered at random in time or space. Suppose there is a collection of mA red objects and nA green objects which are all identical, apart from their colour, and are scattered at random over a region of area A. Focusing on the red objects alone (Diagram **R**) the pattern will form a spatial Poisson process with rate m per unit area. Similarly, the green objects (Diagram **G**) will form a spatial Poisson process with a rate of n per unit area. However, a colour-blind person would see a combined set of objects (Diagram **C**) randomly distributed at an average rate of $(m + n)$ objects per unit area.

Example 9

At breakfast time on a Monday morning, the mean number of greenfinches at a bird feeding station is 2.4. The mean number of other types of finch is 0.14. Assuming that the numbers of finches present are observations from independent Poisson distributions, determine the probability that, on a randomly chosen Monday breakfast time, there are exactly 4 finches at the feeding station.

The total number of finches is the number of greenfinches plus the number of other types of finch, and has a Poisson distribution. The mean number is $2.4 + 0.14 = 2.54$. The required probability is therefore:

$$e^{-2.54}\frac{(2.54)^4}{4!} = 0.137 \text{ (to 3 dp)}$$

Example 10

An observer is standing beside a road. Both cars and lorries pass the observer at random points in time. On average there are 300 cars per hour, while the mean time between lorries is 5 minutes. Determine the probability that exactly 6 vehicles pass the observer in a one-minute period.

Since the question refers to 'random points in time' a Poisson distribution is appropriate both for cars and for lorries. Since the mean time between lorries is 5 minutes, on average 12 lorries pass per hour, so the combined rate is 312 vehicles per hour, which corresponds to $\frac{312}{60} = 5.2$ vehicles per minute. The required probability is therefore

$$e^{-5.2}\frac{(5.2)^6}{6!} = 0.151 \text{ (to 3 dp)}$$

Exercise 2F

1 A city is split into two regions, North and South. In the peaceful North, traffic accidents happen at the rate of one a month. In the congested South, however, traffic accidents happen at the rate of two a week. Assuming independent Poisson distributions for the two regions, use tables to determine the probability that the city experiences more than twelve accidents in a randomly chosen month.

2 A large lawn contains plants of two type of weed. Daisies occur at an average rate of 2.5 per square metre and dandelions at an average rate of 1.25 per square metre. Assuming independence, find the probability that a randomly chosen 4 square metres of lawn contains a total of between 10 and 15 of these two types of weeds.

3 The numbers of emissions per minute from two radioactive objects A and B are independent Poisson random variables with means 0.65 and 0.45 respectively.

Find the probabilities that:

a) in a period of four minutes there are at least three emissions from A,

b) in a period of two minutes there is a total of less than four emissions from A and B together.

4 The number of customers per hour entering a jeweller's shop has a Poisson distribution. For the first hour after opening the mean rate is 0.6 per hour and for the next three hours the mean rate is 1.3 per hour.

Find the probability that there are between 4 and 6 (inclusive) customers entering the shop in the first four hours.

5 The number of strikes per game, obtained by a tenpin bowler, can be modelled by a Poisson distribution with mean λ. For the first game played, $\lambda = 0.2$ and for each subsequent game played, $\lambda = 1.1$.

In a match consisting of three consecutive independent games, calculate the probability that the tenpin bowler will obtain a total of between 3 and 5 strikes, inclusive. *(AQA, 2003)*

6 The number of smoke alarms sold by a store may be modelled by a Poisson distribution with mean 7.5 per day.

a) Find the probability that during a particular day the store sells:
   i) 6 or fewer smoke alarms,
   ii) exactly 8 smoke alarms.

b) Evaluate the standard deviation of the number of smoke alarms sold by the store per day. *(AQA, 2004)*

Summary

You should now be able to ...	Check out
1 Recognise situations where the random variable has a Poisson distribution.	**1** For each of the following cases determine whether the random variable could have a Poisson distribution: a) The weight of apples in a random collection of apples. b) The number of bad apples in a random collection of 20 apples. c) The length of time between accidents at a busy junction. d) The number of accidents per week at a busy junction. e) The number of toadstools in a lawn on a randomly chosen day.

2 Determine a Poisson probability using the formula.	**2** The random variable X has a Poisson distribution with mean 3.17. Determine: a) $P(X=2)$ b) $P(X=0)$ c) $P(X>2)$
3 Determine a Poisson probability using tables.	**3** The random variable X has a Poisson distribution with mean 3. Determine: a) $P(X=2)$ b) $P(X=0)$ c) $P(X>2)$
4 Know the relation between the mean, variance and standard deviation of a Poisson distribution.	**4** A random variable has a Poisson distribution with mean 36. Determine the standard deviation of the distribution.
5 Work with sums of independent Poisson random variables.	**5** The independent random variables $X_1, X_2, \ldots,$ each have a Poisson distribution with mean 1.3. Determine: a) $P(X_1 + X_2 \leqslant 2)$ b) $P\left(\sum_{i=1}^{5} X_i \geqslant 6\right)$ c) $P\left(10 \leqslant \sum_{i=1}^{10} X_i \leqslant 20\right)$

S2

Revision exercise 2

1 A random variable X has a Poisson distribution with mean 2.0. Show that, correct to four decimal places:

a) $P(X = 1) = 0.2707$,

b) $P(X \geqslant 6) = 0.0166$. *(AQA, 2001)*

2 The number of letters of complaint received by a department store follows a Poisson distribution with mean 6.5 per day. Find the probability that on a particular day:

a) 7 or fewer letters of complaint are received,

b) exactly 7 letters of complaint are received. *(AQA, 2001)*

3 The number of vehicles arriving at a toll bridge during a 5-minute period can be modelled by a Poisson distribution with mean 3.6.

a) Calculate:

i) the probability that at least 3 vehicles arrive in a 5-minute period,

ii) the probability that at least 3 vehicles arrive in each of three successive 5-minute periods.

b) Show that the probability that no vehicles arrive in a 10-minute period is 0.0007, correct to four decimal places. *(AQA, 2002)*

4 The number of mobile phone calls, X, that Pat makes each day to her boyfriend James has a Poisson distribution with a mean of 4.0. The number of text messages, Y, that James sends each day to Pat has a Poisson distribution with a mean of 3.5. Assume that X and Y are independent random variables.

a) Calculate the probability that, on a randomly chosen day:
   i) Pat phones James on more than 8 occasions,
   ii) James sends fewer than 2 text messages to Pat.

b) The total number of phone calls made and text messages sent, T, is a Poisson random variable with mean λ.
   i) Write down the value of λ.
   ii) Calculate the probability that, on a randomly selected day, T is at least eleven.

(*AQA, 2004*)

5 The number of cars, travelling from East to West, passing a point on a motorway, may be modelled by a Poisson distribution with a mean of 1.2 per five-second interval.

a) Find the probability that, during a particular five-second interval, the number of cars, travelling from East to West, which pass the point is:
   i) zero, ii) exactly two.

The number of cars, travelling from West to East, passing the same point on the motorway, may be modelled by a Poisson distribution with a mean of 3.8 per five-second interval.

b) Find the probability that, during a particular five-second interval, more than eight cars, travelling from West to East, pass the point.

c) Find the probability that, during a particular five-second interval, the total number of cars passing the point is less than eight.

(You may assume that the number of cars travelling from East to West is independent of the number travelling from West to East.)

d) Explain why a Poisson distribution may not provide an adequate model for the total number of car passengers passing the point in a five-second interval.

(*AQA, 2001*)

6 The number of cyclists passing a point on a cycle track alongside a main road into Oxford, during the morning rush hour, may be modelled by a Poisson distribution with mean 2.8 per minute.

a) Find the probability that the number of cyclists passing the point during a particular minute in the morning rush hour is:
   i) one or fewer, ii) three or more.

b) Find the probability that fewer than ten cyclists pass the point during a **five**-minute interval in the morning rush hour.

c) The numbers of cyclists passing the point during ten, randomly selected, one-minute intervals over a twenty-four hour period were:

   0 0 5 1 0 4 1 0 0 2

   i) Calculate the mean and variance of this sample.
   ii) Give a reason, based on your results in part c) i), why the Poisson distribution is unlikely to provide an adequate model for the number of cyclists passing the point over a twenty-four hour period.
   iii) Give a reason, **not** based on your results in part c) i), why the Poisson distribution is unlikely to provide an adequate model for the number of cyclists passing the point over a twenty-four hour period.

(*AQA, 2002*)

7 Bronwen runs a post office in a large village. The number of registered letters posted at this office may be modelled by a Poisson distribution with mean 1.4 per day.

a) Find the probability that at this post office:
   i) two or fewer registered letters are posted on a particular day;
   ii) a total of four or more registered letters are posted on two consecutive days.

b) The village also contains a post office run by Gopal. Here the number of registered letters posted may be modelled by a Poisson distribution with mean 2.4 per day.

   Find the probability that, on a particular day, the number of registered letters posted at Gopal's post office is less than 4.

c) The numbers of registered letters posted at the two post offices are independent. Find the probability that, on a particular day, the total number of registered letters posted at the two post offices is more than six.

d) Give one reason why the Poisson distribution might **not** provide a suitable model for the number of registered letters posted daily at a post office. (*AQA, 2002*)

8 Karen is an engineer who is responsible for maintaining a textile machine during the night shift. She is called when the machine operator believes that the machine needs adjustment or repair.

a) The number of times Karen is called during a night shift may be modelled by a Poisson distribution with mean 0.8. Find the probability that, on a particular night shift, she is called:
   i) two or fewer times, ii) at least once, iii) exactly once.

b) Karen is now made responsible for 5 machines. These machines are operated independently. For each machine, the number of times she is called during a night shift may be modelled by a Poisson distribution with mean 0.8.
   i) State the distribution of the total number of times she is called during a night shift.
   ii) Find the probability that, during a night shift, she is called 5 or more times.
   iii) Find the probability that, during a night shift, she is called at least once to each of the 5 machines. (*AQA, 2003*)

9 The number of family groups entering a museum on an August afternoon may be modelled by a Poisson distribution with mean 0.60 per minute.

a) Find the probability that during a particular minute 2 or fewer such groups enter the museum.

b) Find the probability that during a particular **five-minute** interval more than 4 such groups enter the museum.

c) Explain why a Poisson distribution would not provide a suitable model for the number of **visitors** entering the museum during one-minute intervals on an August afternoon. (*AQA, 2003*)

3 Continuous random variables

This chapter will show you how to

- Understand the differences between discrete and continuous random variables
- Determine a distribution function from a probability density function, and vice-versa
- Determine the probability of a continuous random variable lying in a specified interval
- Determine the median, quartiles and percentiles of a continuous random variable
- Determine the mean, variance and standard deviation of a continuous random variable and of simple functions of a continuous random variable
- Recognise a rectangular distribution and be able to determine its properties

Before you start

You should know how to ...	Check in
1 Summarise data using classes.	**1** Summarize the following data, using the classes 0–, 10–, 20–, 30–50: 5 15 23 34 4 32 45 8 18 22 22 44 49 21 16 18 19 2 26 31
2 Display grouped data using a histogram, with possibly unequal class widths.	**2** Display the following data using a histogram: x: 0– 20– 30– 40– 50–100 f: 12 13 33 27 15
3 Understand what is meant by the median and quartiles.	**3** Determine the median and quartiles of the data given in question **1** above.
4 Differentiate a polynomial.	**4** Differentiate the following expressions with respect to x: a) x^2 b) $3x^3 - 2x + 1$.
5 Calculate simple definite integrals.	**5** a) Evaluate each of the following integrals: i) $\int_1^3 x\,\mathrm{d}x$ ii) $\int_0^4 x^2\,\mathrm{d}x$ iii) $\int_{-1}^1 (x^3 + 4x^2 - 8x)\,\mathrm{d}x$ b) Express the following as simple functions of x: i) $\int_0^x t\,\mathrm{d}t$ ii) $\int_{-1}^x t^2\,\mathrm{d}t$ iii) $\int_x^1 (t^3 - 2t)\,\mathrm{d}t$

3.1 Differences from discrete random variables

Chapters 1 and 2 have focused on discrete random variables, quantities whose values may be unpredictable but for which a list of the possible values can be made. **Continuous random variables** differ in that no such list can be made, though the range of values can be described. For a continuous random variable, the number of possible values is infinite. Here are some examples:

Continuous random variable	Possible values
The height of a randomly chosen 18-year old male student	1.3 m to 2.3 m (say)
The true weight of a '1 kg' bag of sugar	990 g to 1010 g
The time between successive earthquakes of magnitude > 7 on the Richter scale	Any length of time

S2

All the above are **measurements** of **physical** quantities. This contrasts with the random variables of the previous chapters, which were mostly concerned with counts.

Measurements of length, time, weight, etc. are continuous random variables.

With continuous random variables, the number of values is limited only by the inefficiency of our measuring instruments. For a continuous random variable (as will be shown later) the probability of any particular exact value is zero – instead, probabilities are associated with ranges of values.

For a continuous random variable: $P(X < x) = P(X \leqslant x)$.

Displaying data on continuous random variables

The appropriate method for displaying data on a continuous random variable is a histogram, in which the areas of the sections represent frequencies, and the heights represent relative frequency densities.

As an illustration, consider some data concerning the geyser known as 'Old Faithful', which is situated in Yellowstone National Park in Wyoming, USA. This geyser is a great tourist attraction because of the regularity of its eruptions of steam. In August 1985, the geyser was watched continuously for a fortnight, with the times between its eruptions being recorded to the nearest minute. The first 50 times are shown below.

This data set was introduced in Chapter 4 of Statistics S1.

80	71	57	80	75	77	60	86	77	56
81	50	89	54	90	73	60	83	65	82
84	54	85	58	79	57	88	68	76	78
74	85	75	65	76	58	91	50	87	48
93	54	86	53	78	52	83	60	87	49

Using class boundaries at 39.5, 49.5, ..., just two observations fall in the 39.5–49.5 class, so the relative frequency for that class is $\frac{2}{50} = 0.04$. Since the class is 10 minutes wide, on a per minute basis this is a relative frequency density of $\frac{0.04}{10} = 0.004$ per minute.

The histogram based on relative frequency densities looks like this:

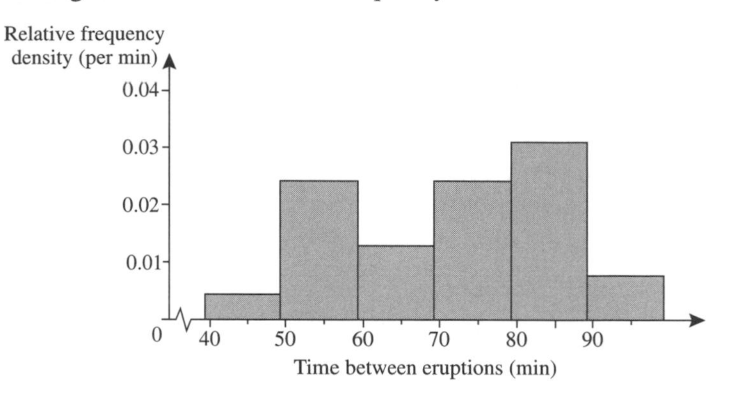

As the sample size increases and smaller class-widths are used so the outline becomes ever smoother. For the 'Old Faithful' data it would ultimately look something like this:

S2

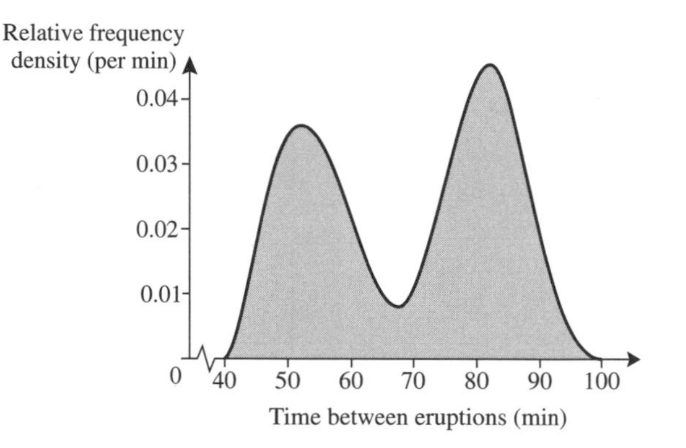

A fuller description of the effects of increasing the sample size is given in Statistics S1.

3.2 Probability density functions, distribution functions and their relationship

The probability density function, f

The data from 'Old Faithful' suggests a general result: as the sample size increases (with correspondingly narrower class intervals), the outline of a histogram will usually converge on a smooth curve. The areas of the individual sections of a histogram represent relative frequencies. As the sample size increases, so sample relative frequencies approach the corresponding population probabilities. Similarly, relative frequency densities become probability densities. The resulting graph is then referred to as the graph of the **probability density function** and the function is denoted by f.

'probability density function' is often abbreviated to **pdf**.

The relative frequency density of a value between 50 and 60 was represented in the previous diagram by a rectangle. In the limit, as relative frequency density becomes probability density, so the corresponding region under the probability density function represents probability.

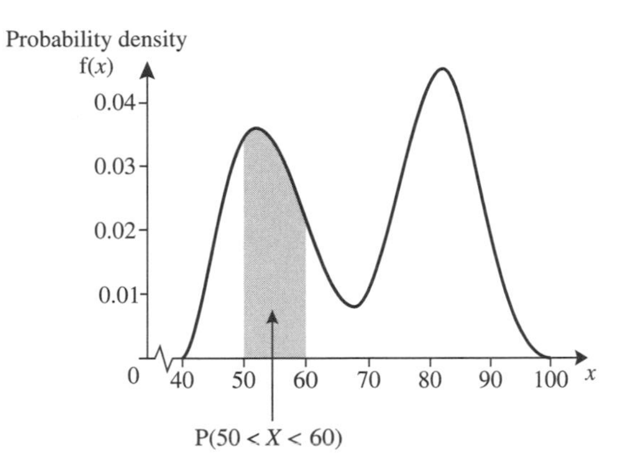

Thus $P(c < X < d)$ is the area of the section between c and d under the graph of f.

S2

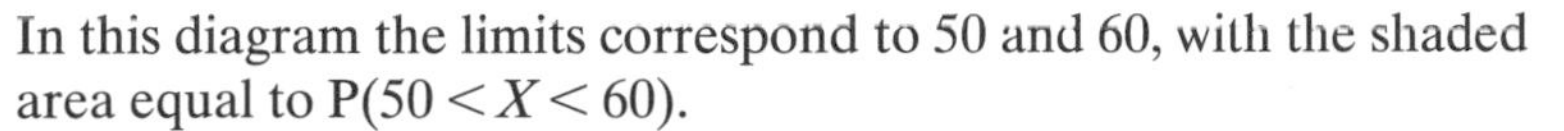

In this diagram the limits correspond to 50 and 60, with the shaded area equal to $P(50 < X < 60)$.

Note that the function f measures probability *density*, not probability.

If the lower limit, 50, is replaced by the minimum possible value for X, then the resulting shaded area is equal to $P(X < 60)$:

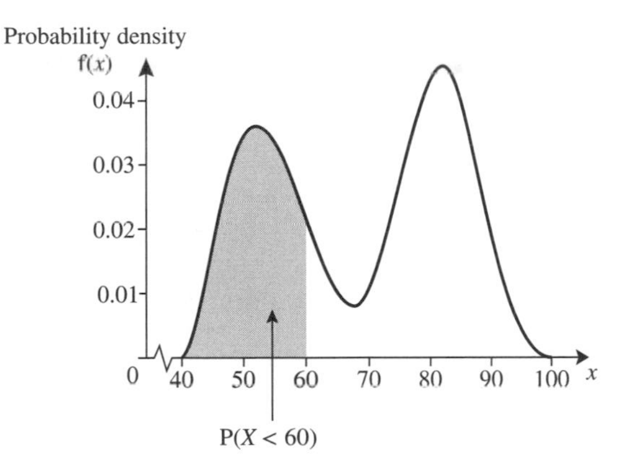

Similarly, if 60 is replaced by the maximum possible value for X, then the resulting area is equal to $P(X > 50)$:

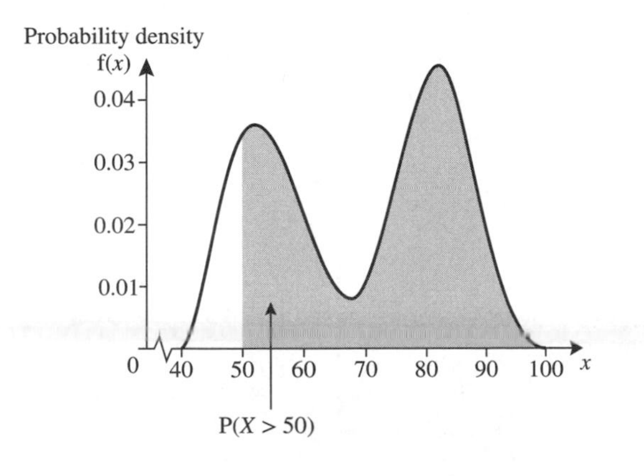

Since the value of X must lie between the least and greatest possible values, the total area under the graph of the probability density function is equal to 1.

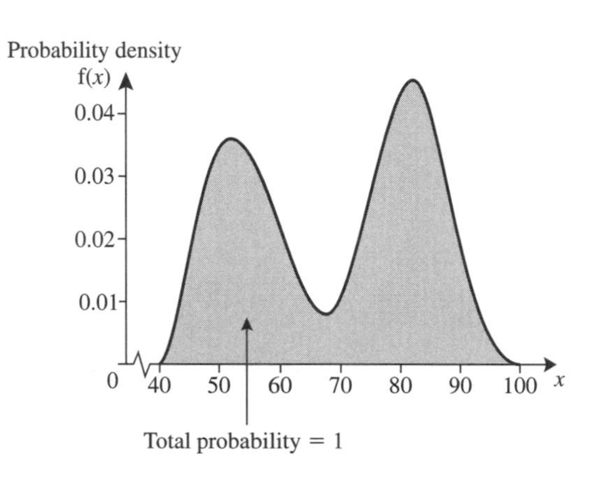

Since negative probabilities are impossible, the graph of f cannot dip below the x-axis:

$$f(x) \geqslant 0 \quad \text{for all values of } x \tag{3.1}$$

The distribution function, F

The distribution function, F, is defined by:

$$F(x) = P(X \leqslant x) = P(X < x) \tag{3.2}$$

The function F is also called the **cumulative distribution function**, abbreviated to **cdf**.

Thus:

$$P(X \leqslant 60) = F(60)$$
$$P(X > 50) = 1 - P(X \leqslant 50) = 1 - F(50)$$

Since an area under a graph can be expressed as an integral, it follows that F is related to the probability density function f by:

$$F(x) = \int_{-\infty}^{x} f(t)\, dt \tag{3.3}$$

You can replace t by any letter, without changing the value of the integral.

The lower limit of the integral is given as $-\infty$, but is in effect the smallest possible value of X.

Since it is always impossible to have a value of X smaller than $-\infty$ and it is certain that any value of X is less than ∞,

$$F(-\infty) = 0 \qquad F(\infty) = 1$$

Strictly $F(-\infty)$ means 'the limiting value of $F(x)$ as x approaches $-\infty$' and $F(\infty)$ is similarly defined.

Here are illustrations of the functions f and F for two examples where X only takes values between a and b, so that $P(X < a) = 0$ and $P(X > b) = 0$.

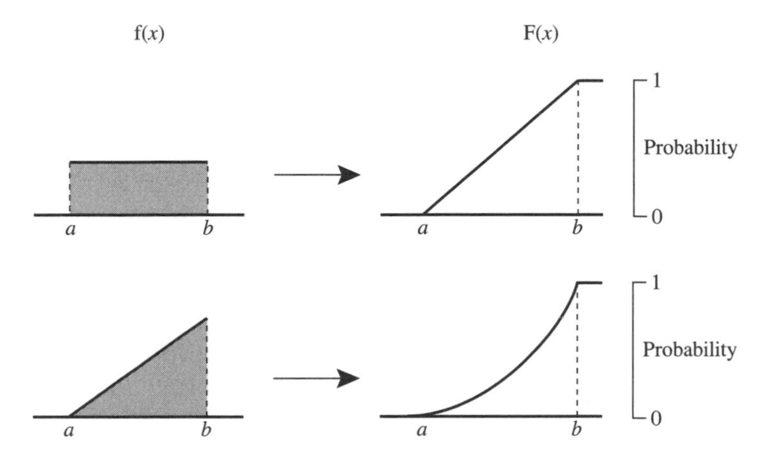

As x increases so $F(x)$, which is a continuous function, either increases or remains constant, but never decreases.

S2

Example 1

The continuous random variable X has probability density function, f, given by:

$$f(x) = \begin{cases} 1 & 2 < x < 3 \\ 0 & \text{otherwise} \end{cases}$$

a) Sketch the graph of f.

b) Determine the form of $F(x)$.

c) Sketch the graph of F.

The statement that $f(x) = 0$ 'otherwise' merely emphasises that attention may safely be restricted to the interval $2 < x < 3$ or, equivalently, to $2 \leqslant x \leqslant 3$.

a) f(x) 1 0 2 3 x

b) Since X cannot have a value less than 2, $F(x) = 0$ for $x \leqslant 2$. For $2 \leqslant x \leqslant 3$, by definition:

$$F(x) = \int_{-\infty}^{x} 1\,dt$$

However, since $f(x) = 0$ for $x \leqslant 2$, the lower limit can be set to 2. Thus:

$$\begin{aligned} F(x) &= \int_{2}^{x} 1\,dt \\ &= [t]_2^x \\ &= (x-2) \end{aligned}$$

Since X cannot have a value greater than 3, $F(x) = 1$ for $x \geqslant 3$. The form of $F(x)$ is therefore:

$$F(x) = \begin{cases} 0 & x \leqslant 2 \\ x - 2 & 2 \leqslant x \leqslant 3 \\ 1 & x \geqslant 3 \end{cases}$$

> Using '$\leqslant$' means that there are apparently two definitions for F(2) and F(3). However, in each case, the two definitions give the same values (namely, $F(2) = 0$ and $F(3) = 1$) emphasizing that $F(x)$ is continuous.

c)

S2

Example 2

The continuous random variable X has probability density function, f, given by;

$$f(x) = \begin{cases} \dfrac{a}{27}x^2 & 0 < x < a \\ 0 & \text{otherwise} \end{cases}$$

where a is a constant.

a) Sketch the graph of f.

b) Determine the form of $F(x)$ and the value of a.

c) Sketch the graph of F.

a)

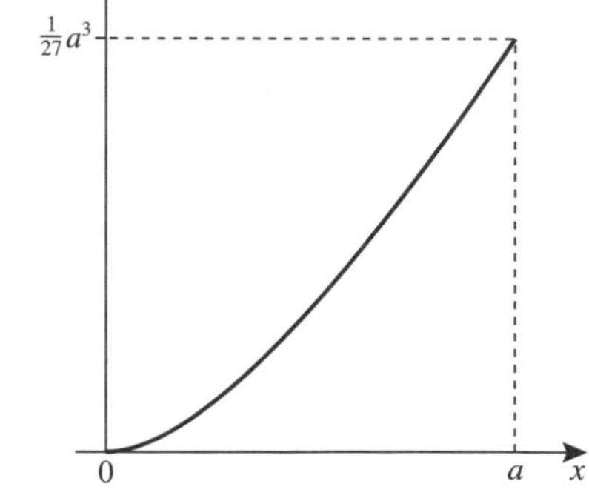

b) For $0 \leqslant x \leqslant a$:

$$\begin{aligned} F(x) &= \frac{a}{27}\int_0^x t^2 \, dt \\ &= \frac{a}{27}\left[\frac{1}{3}t^3\right]_0^x \\ &= \frac{a}{81}x^3 \end{aligned}$$

Since X does not take values greater than a, $\mathrm{F}(a) = 1$.
Thus a satisfies:

$$\frac{a}{81} \times a^3 = \frac{a^4}{81} = 1$$

Since $3^4 = 81$, it follows that $a = 3$. Thus the full description of $\mathrm{F}(x)$ is:

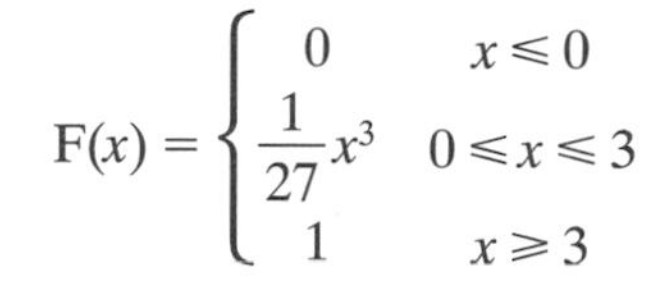

$$\mathrm{F}(x) = \begin{cases} 0 & x \leqslant 0 \\ \dfrac{1}{27}x^3 & 0 \leqslant x \leqslant 3 \\ 1 & x \geqslant 3 \end{cases}$$

An equivalent procedure for finding the value of a is to solve $\int_0^a \mathrm{f}(x)\,\mathrm{d}x = 1$.

c)

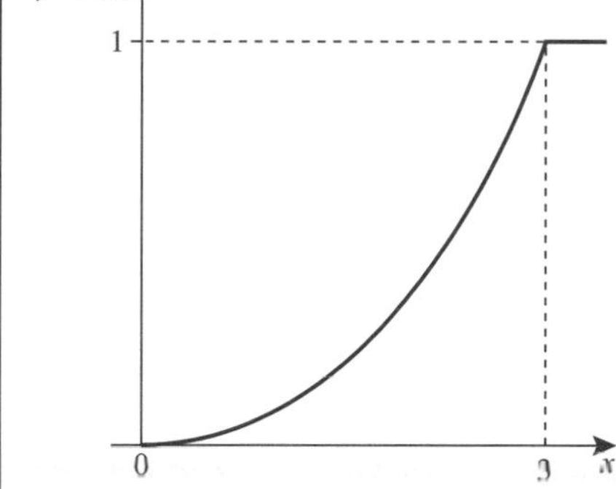

S2

Calculator practice

If you have a graphical calculator then you should use it to check your sketch of the graph of F in Example 2.

Problems involving probability density functions often require the calculation of areas. When the graph of f is made up of straight-line sections, it is often much quicker to use standard geometric results.

Sometimes the probability density function may be defined 'piece-wise', as in the next example.

Example 3

The continuous random variable X has probability density function given by:

$$\mathrm{f}(x) = \begin{cases} 2k(x-1) & 1 < x < 2 \\ k(4-x) & 2 < x < 4 \\ 0 & \text{otherwise} \end{cases}$$

where k is a positive constant.

a) Sketch the graph of f.
b) Determine the value of k.
c) Determine the form of $\mathrm{F}(x)$.
d) Sketch the graph of F.

a) 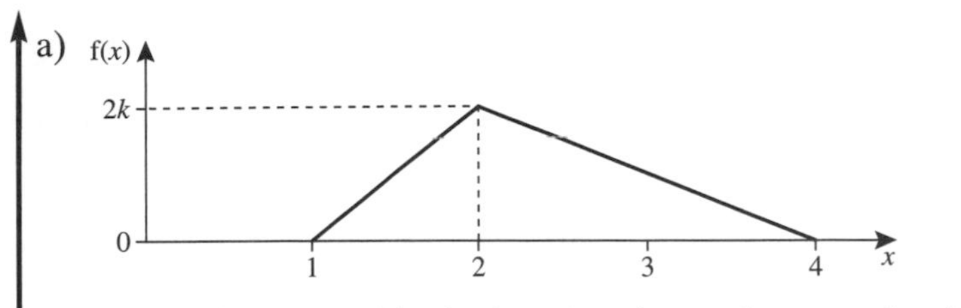

b) To find the value of k, the fact that the total area under the graph is equal to 1 is used.

The sketch reveals that the region of interest is a triangle with height equal to $2k$ and base equal to $(4 - 1) = 3$. The triangle has area $\frac{1}{2}(3 \times 2k) = 3k$. Setting this area equal to 1 gives $k = \frac{1}{3}$.

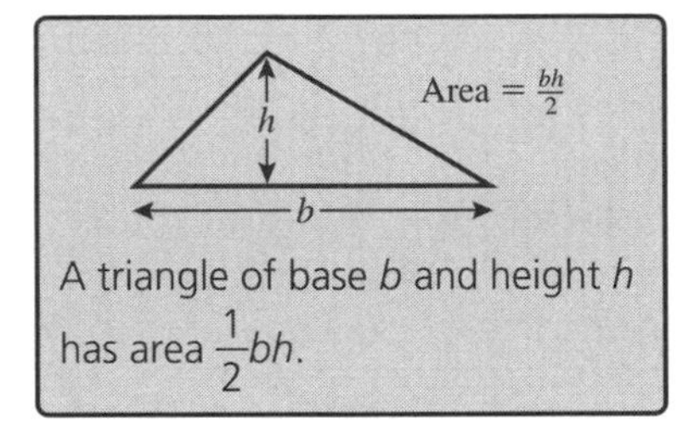

A triangle of base b and height h has area $\frac{1}{2}bh$.

S2

c) $F(x) = 0$ for $x \leqslant 1$ and $F(x) = 1$ for $x \geqslant 4$, since X only takes values in the interval 1 to 4. The two intervals $(1 < x < 2)$ and $(2 < x < 4)$ are considered separately since f has a different form in each interval.

For $1 \leqslant x \leqslant 2$:

$$\begin{aligned} F(x) &= \int_1^x 2k(t-1)\,dt \\ &= 2k\left[\frac{1}{2}(t-1)^2\right]_1^x \\ &= k(x-1)^2 \\ &= \frac{1}{3}(x-1)^2 \end{aligned}$$

In particular, $P(X \leqslant 2) = F(2) = \frac{1}{3}$.

For $2 \leqslant x \leqslant 4$:

$$\begin{aligned} F(x) &= \int_1^x f(t)\,dt \\ &= \int_1^2 f(t)dt + \int_2^x f(t)dt \\ &= F(2) + \int_2^x k(4-t)dt \\ &= \frac{1}{3} + k\left[-\frac{1}{2}(4-t)^2\right]_2^x \\ &= \frac{1}{3} - \frac{k}{2}\{(4-x)^2 - (4-2)^2\} \\ &= \frac{1}{3} + \frac{1}{6}\{4 - (4-x)^2\} \\ &= \frac{2 + 4 - (4-x)^2}{6} \\ &= 1 - \frac{1}{6}(4-x)^2 \end{aligned}$$

Note that, for $2 \leqslant x \leqslant 4$, $F(x)$ is NOT $\int_2^x f(t)dt$; it is $F(2) + \int_2^x f(t)dt$.

As a check, note that F(4) does indeed equal the maximum value, 1. Also, setting $x = 2$,

$$\frac{1}{3}(2-1)^2 = \frac{1}{3} = 1 - \frac{1}{6}(4-2)^2$$

as required by the continuity of F.

The complete description of F(x) is therefore:

$$\text{F}(x) = \begin{cases} 0 & x \leqslant 1 \\ \frac{1}{3}(x-1)^2 & 1 \leqslant x \leqslant 2 \\ 1 - \frac{1}{6}(4-x)^2 & 2 \leqslant x \leqslant 4 \\ 1 & x \geqslant 4 \end{cases}$$

d)

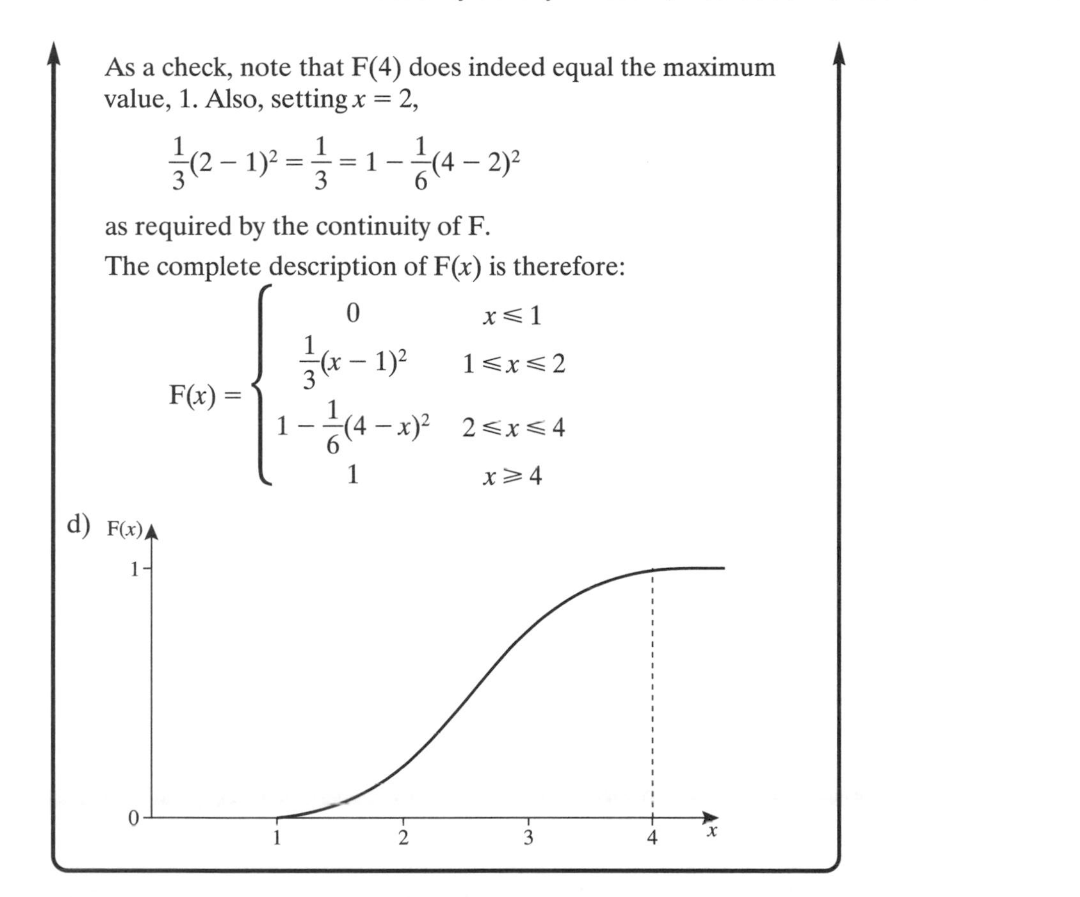

S2

Calculator practice

If you have a graphical calculator then you should use it to check the graph of F in Example 3, for $1 \leqslant x \leqslant 2$ and $2 \leqslant x \leqslant 4$.

Exercise 3A

For each of Questions 1 to 5, the random variable X has probability density function f and distribution function F.

a) Sketch the graph of f.

b) Determine F(x).

c) Sketch the graph of F.

1 $\text{f}(x) = \begin{cases} \frac{1}{8}(2+x) & 1 < x < 3 \\ 0 & \text{otherwise} \end{cases}$

2 $\text{f}(x) = \begin{cases} -\frac{1}{4}x & -2 < x < 0 \\ \frac{1}{4}x & 0 < x < 2 \\ 0 & \text{otherwise} \end{cases}$

3 $f(x) = \begin{cases} \frac{1}{2} & 1 < x < 2 \\ \frac{1}{4} & 2 < x < 4 \\ 0 & \text{otherwise} \end{cases}$

4 $f(x) = \begin{cases} \frac{1}{3}x^2 & -2 < x < 1 \\ 0 & \text{otherwise} \end{cases}$

5 $f(x) = \begin{cases} \frac{3}{19}(x+2)^2 & 0 < x < 1 \\ 0 & \text{otherwise} \end{cases}$

S2

6 For each of the following functions, state, giving a reason, whether or not there is a value of the constant k for which the function can be a probability density function. Sketch graphs may help.

a) $f(x) = \begin{cases} kx & -1 < x < 1 \\ 0 & \text{otherwise} \end{cases}$

b) $f(x) = \begin{cases} kx^2 & -1 < x < 2 \\ 0 & \text{otherwise} \end{cases}$

c) $f(x) = \begin{cases} 1 + kx & 0 < x < 3 \\ 0 & \text{otherwise} \end{cases}$

d) $f(x) = \begin{cases} 4 + x^2 & -k < x < k \\ 0 & \text{otherwise} \end{cases}$

7 It is given that:

$$f(x) = \begin{cases} kx^2 - c & 1 < x < 2 \\ 0 & \text{otherwise} \end{cases}$$

where k and c are positive constants.

a) Sketch the graph of f.

b) Find, in terms of k, the maximum possible value for c.

c) With this value for c, find the corresponding value of k.

d) With these values for k and c, find $F(x)$.

Obtaining f from F

Since F can be obtained by integrating f, it follows that f can be obtained by differentiating F:

$$f(x) = \frac{d}{dx}F(x)$$

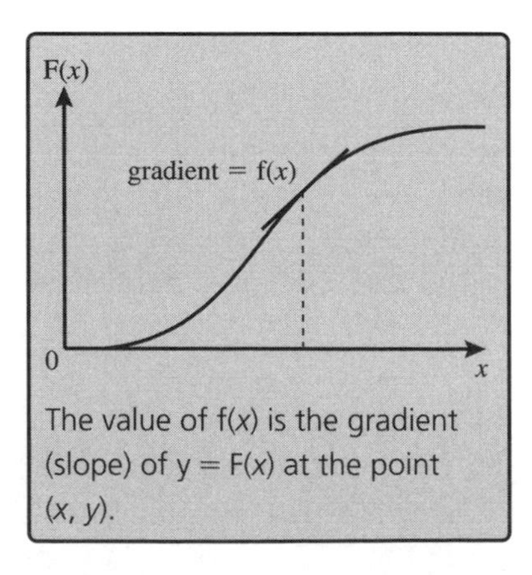

The value of f(x) is the gradient (slope) of y = F(x) at the point (x, y).

Example 4

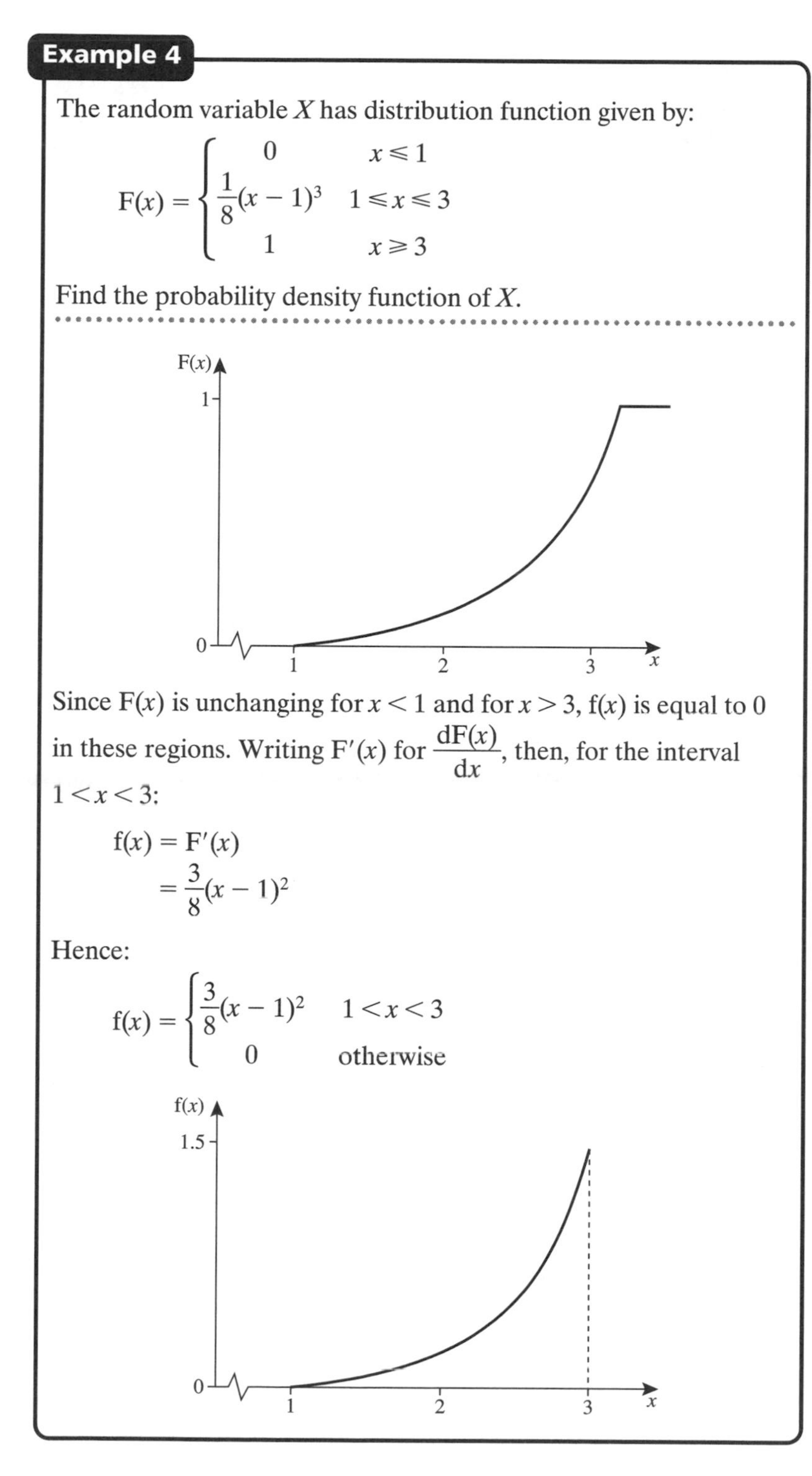

The random variable X has distribution function given by:

$$F(x) = \begin{cases} 0 & x \leqslant 1 \\ \frac{1}{8}(x-1)^3 & 1 \leqslant x \leqslant 3 \\ 1 & x \geqslant 3 \end{cases}$$

Find the probability density function of X.

Since $F(x)$ is unchanging for $x < 1$ and for $x > 3$, $f(x)$ is equal to 0 in these regions. Writing $F'(x)$ for $\frac{dF(x)}{dx}$, then, for the interval $1 < x < 3$:

$$\begin{aligned} f(x) &= F'(x) \\ &= \frac{3}{8}(x-1)^2 \end{aligned}$$

Hence:

$$f(x) = \begin{cases} \frac{3}{8}(x-1)^2 & 1 < x < 3 \\ 0 & \text{otherwise} \end{cases}$$

Strictly, $F(x)$ does not have a derivative at $x = 3$, so $f(3)$ cannot be determined. It is conventional to use '0 otherwise' to cover this situation.

S2

Exercise 3B

In Questions 1 to 5 the probability density function and distribution function of the continuous random variable X arc f and F respectively.

1 It is given that:

$$F(x) = \begin{cases} 0 & x \leqslant 1 \\ x - 1 & 1 \leqslant x \leqslant 2 \\ 1 & x \geqslant 2 \end{cases}$$

Find f(x) and sketch the graph of f.

2 It is given that:

$$F(x) = \begin{cases} 0 & x \leqslant 0 \\ ax^2 & 0 \leqslant x \leqslant 4 \\ 1 & x \geqslant 4 \end{cases}$$

Find: a) the value of the constant a

b) f(x)

3 It is given that:

$$F(x) = \begin{cases} 0 & x \leqslant 1 \\ a + bx^2 & 1 \leqslant x \leqslant 3 \\ 1 & x \geqslant 3 \end{cases}$$

a) Find the values of the constants a and b.

b) Find f(x) and sketch the graph of f.

4 It is given that:

$$F(x) = \begin{cases} 0 & x \leqslant 4 \\ \frac{1}{4}(x - 4) & 4 \leqslant x \leqslant 8 \\ 1 & x \geqslant 8 \end{cases}$$

Find f(x) and sketch the graph of f.

5 It is given that:

$$F(x) = \begin{cases} 0 & x \leqslant 0 \\ \frac{1}{8}x^2 & 0 \leqslant x \leqslant 2 \\ a + b(4 - x)^2 & 2 \leqslant x \leqslant 4 \\ 1 & x \geqslant 4 \end{cases}$$

a) Find the values of the constants a and b.

b) Find f(x).

c) Sketch the graph of f.

3.3 The probability of an observation lying in a specified interval

In Section 3.2 (p 53) you saw that the area of a section under the graph of the probability density function represents a probability.

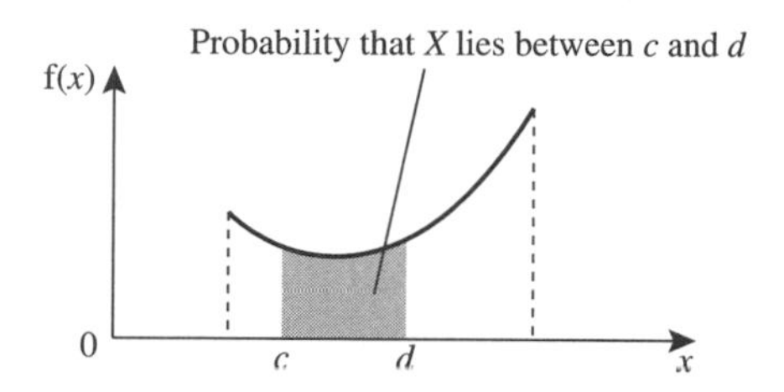

The probability of X taking a value in the interval $(c < x < d)$ is given by the corresponding area. Since the area between the graph of f and the x-axis is given by the integral of $f(x)$ with respect to x, then

$$P(c < X < d) = \int_c^d f(x)\,dx \qquad (3.4)$$

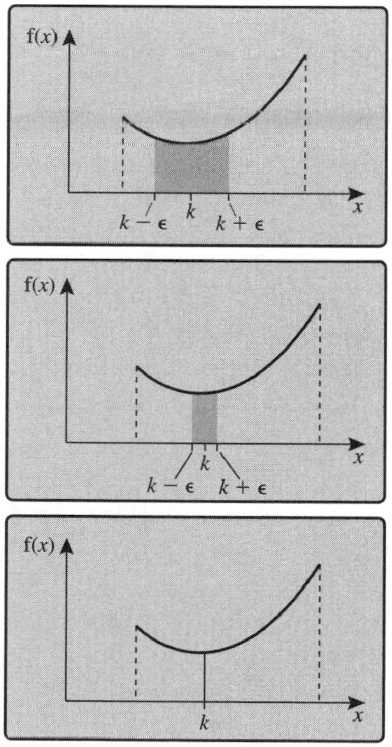

Equation (3.4) applies for all c and d. Setting $c = k - \epsilon$ and $d = k + \epsilon$ therefore gives:

$$P(k - \epsilon < X < k + \epsilon) = \int_{k-\epsilon}^{k+\epsilon} f(x)dx$$

As ϵ approaches 0, the left-hand side approaches $P(X = k)$ and the right-hand side approaches $\int_k^k f(x)dx$ which is zero.

Thus, for any value, k:

$$P(X = k) = 0$$

This is an entirely general result for any **continuous** random variable and implies that you need not be fussy about whether you write $P(X < x)$ or $P(X \leqslant x)$, since (as claimed previously):

$$P(X < x) = P(X \leqslant x) \quad \text{for all } x$$

The total of a set of relative frequencies is, by definition, equal to 1. The same is true for probabilities. The total area between the graph of $f(x)$ and the x-axis is therefore 1. For a random variable that can only take values between a and b:

$$\int_a^b f(x)dx = 1 \qquad (3.5)$$

as illustrated in the diagram.

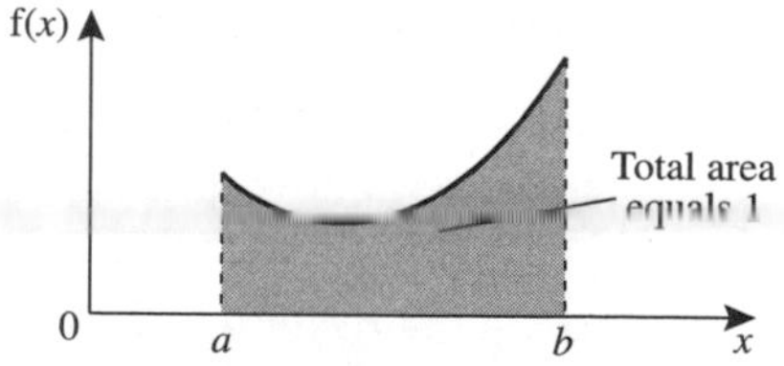

Although f(x) often has values less than 1, this need not be the case. For example:

$$f(x) = \begin{cases} 2 & 0 < x < \frac{1}{2} \\ 0 & \text{otherwise} \end{cases}$$

defines a proper probability density function with total area 1.

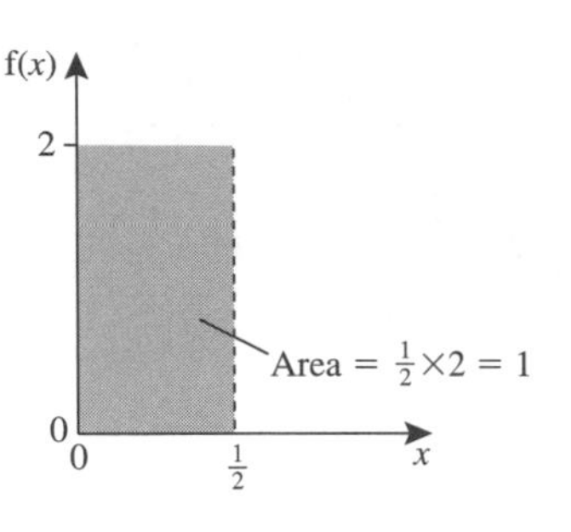

Example 5

The continuous random variable X has probability density function given by

$$f(x) = \begin{cases} \frac{1}{2}x & 0 < x < 2 \\ 0 & \text{otherwise} \end{cases}$$

S2

a) Sketch the graph of f. b) Determine $P(X > 1)$.

a)

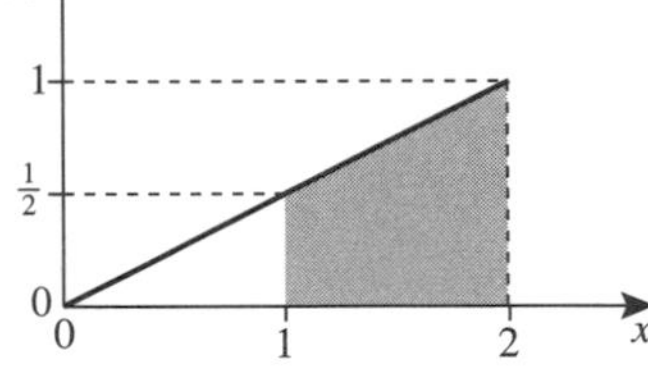

Sketching graphs of functions is covered in the C1 module.

b) The diagram shows that the area corresponding to $P(X > 1)$ is greater than half of the total area between f(x) and the x-axis, so that the required probability will be greater than 0.5. If the calculations give a value smaller than 0.5 then there must be an error (possibly in the diagram!).

Method 1: Calculus

The required probability is:

$$\int_1^2 \frac{x}{2}\,dx = \left[\frac{x^2}{4}\right]_1^2$$

$$= \frac{4-1}{4}$$

$$= \frac{3}{4}$$

So $P(X > 1) = 0.75$, which, as anticipated, is considerably greater than 0.5.

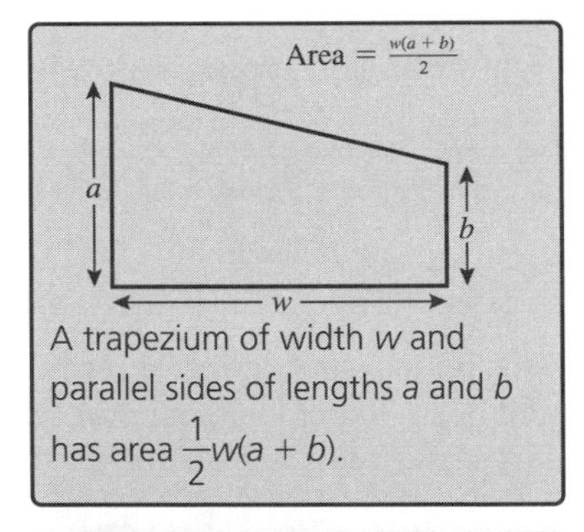

A trapezium of width w and parallel sides of lengths a and b has area $\frac{1}{2}w(a + b)$.

Method 2: Geometry

The required probability is given by the area of a trapezium having parallel sides of lengths $\frac{1}{2}$ and 1 at a distance apart $(2 - 1) = 1$.

The area corresponding to $P(X > 1)$ is therefore equal to:

$$\frac{1}{2} \times \left(\frac{1}{2} + 1\right) = \frac{3}{4}$$

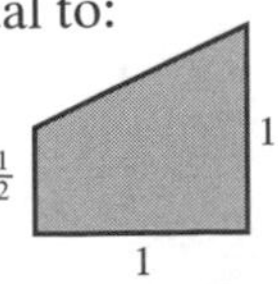

confirming the result found by calculus.

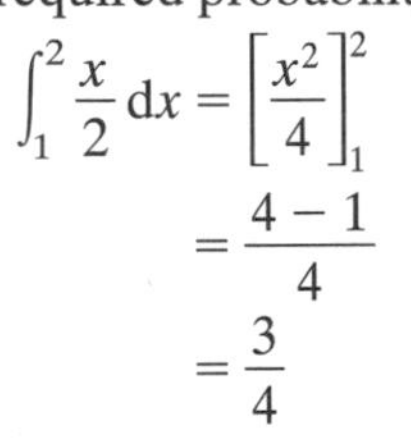

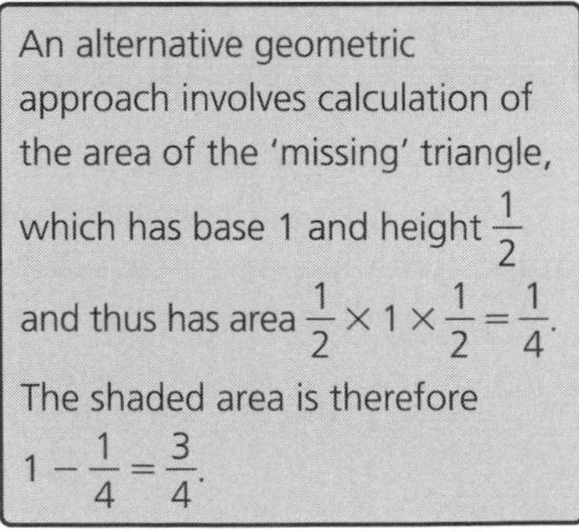

An alternative geometric approach involves calculation of the area of the 'missing' triangle, which has base 1 and height $\frac{1}{2}$ and thus has area $\frac{1}{2} \times 1 \times \frac{1}{2} = \frac{1}{4}$.

The shaded area is therefore $1 - \frac{1}{4} = \frac{3}{4}$.

Example 6

The continuous random variable Y has probability density function given by:

$$f(y) = \begin{cases} |y-1| & 0 < y < 2 \\ 0 & \text{otherwise} \end{cases}$$

a) Sketch the graph of f.

b) Determine $P\left(\frac{1}{3} < Y < \frac{4}{3}\right)$

---

a) First the formula for the probability density function is rewritten so that the '|' signs are not needed:

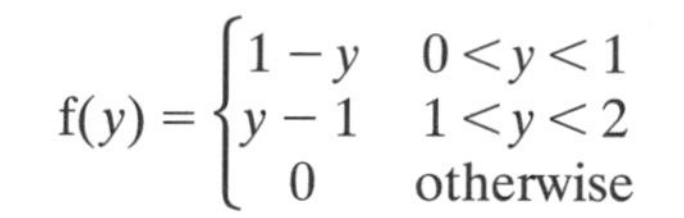

$$f(y) = \begin{cases} 1-y & 0 < y < 1 \\ y-1 & 1 < y < 2 \\ 0 & \text{otherwise} \end{cases}$$

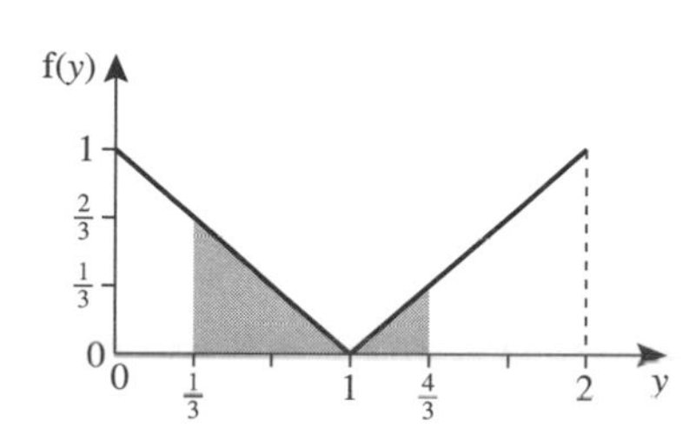

b) The diagram suggests that the answer will be less than 0.5.

Method 1: Calculus

The result that:

$$P\left(\frac{1}{3} < Y < \frac{4}{3}\right) = P\left(\frac{1}{3} < Y < 1\right) + P\left(1 < Y < \frac{4}{3}\right)$$

can be used, giving:

$$\int_{\frac{1}{3}}^{1}(1-y)\,dy + \int_{1}^{\frac{4}{3}}(y-1)\,dy = \left[y - \frac{1}{2}y^2\right]_{\frac{1}{3}}^{1} + \left[\frac{1}{2}y^2 - y\right]_{1}^{\frac{4}{3}}$$

$$= \left\{\left(1 - \frac{1}{2}\right) - \left(\frac{1}{3} - \frac{1}{18}\right)\right\} + \left\{\left(\frac{16}{18} - \frac{4}{3}\right) - \left(\frac{1}{2} - 1\right)\right\}$$

$$= \left(\frac{1}{2} - \frac{5}{18}\right) + \left(\frac{-8}{18} - \frac{-1}{2}\right)$$

$$= \frac{4}{18} + \frac{1}{18}$$

$$= \frac{5}{18}$$

The probability that Y takes a value between $\frac{1}{3}$ and $\frac{4}{3}$ is $\frac{5}{18}$, or 0.278 (to 3 dp).

Remember that:

$$|w| = \begin{cases} w & w \geqslant 0 \\ -w & w \leqslant 0 \end{cases}$$

Thus $|3| = 3$ and $|-4| = 4$.

S2

Method 2: Geometry

In this case there are two triangles. At $y = \frac{1}{3}$, $f(y) = \frac{2}{3}$, so the left-hand triangle has both height and base equal to $\frac{2}{3}$ and therefore has area equal to:

$$\frac{1}{2}\left(\frac{2}{3} \times \frac{2}{3}\right) = \frac{2}{9}$$

At $y = \frac{4}{3}$, $f(y) = \frac{1}{3}$, so the right-hand triangle has area equal to:

$$\frac{1}{2}\left(\frac{1}{3} \times \frac{1}{3}\right) = \frac{1}{18}$$

The total area of the two triangles is therefore $\frac{2}{9} + \frac{1}{18} = \frac{5}{18}$, which agrees with the answer obtained by calculus.

In this case the geometric approach is much simpler than using calculus.

S2

Example 7

The continuous random variable X has probability density function given by:

$$f(x) = \begin{cases} kx^2 & 1 < x < 3 \\ 0 & \text{otherwise} \end{cases}$$

where k is a positive constant.

a) Sketch the graph of f.

b) Determine the value of k.

c) Determine $P(X < 2)$.

a)

b) A geometric solution is not appropriate in this case, since a curve is involved.

The constant k is found by using the fact that the integral of f is 1:

$$\int_1^3 kx^2\,dx = \left[\frac{1}{3}kx^3\right]_1^3 = \frac{1}{3}k(27 - 1)$$
$$= \frac{26}{3}k$$

Since the integral is equal to 1, it follows that $k = \frac{3}{26}$.

c) $$\begin{aligned}P(X<2) &= \int_1^2 kx^2\,dx \\ &= \left[\frac{1}{3}kx^3\right]_1^2 \\ &= \frac{1}{3}k(8-1) \\ &= \frac{7}{3}k \\ &= \frac{7}{26}\end{aligned}$$

The probability that X takes a value less than 2 is $\frac{7}{26}$, or 0.269 (to 3 dp).

S2

Calculator practice

If you have a calculator that can perform numerical integration, then you should check that you know the appropriate instructions. Practice by checking that (to 3 dp) the answers to the previous example are correct.

Previous examples have obtained $P(c < X < d)$ using integrals of the probability density function. Another method is to use the distribution function, F. Since F is defined in terms of cumulative probabilities, two useful relations are:

$$P(c < X < d) = F(d) - F(c) \qquad (3.6)$$

$$P(X > d) = 1 - F(d) \qquad (3.7)$$

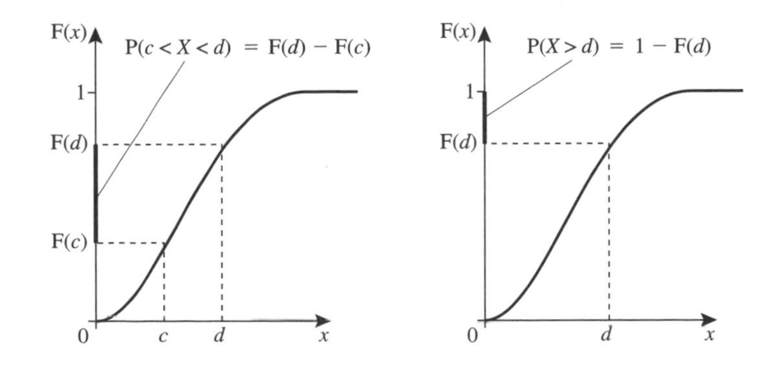

Example 8

The continuous random variable X has probability density function given by

$$f(x) = \begin{cases} kx + \frac{1}{2} & 1 < x < 2 \\ 0 & \text{otherwise} \end{cases}$$

where k is a positive constant.

a) Sketch the graph of f.

b) Determine, in terms of k, a formula for F(x).

c) Hence determine the value of k.

S2

d) Determine P($X > 1.8$).

a)

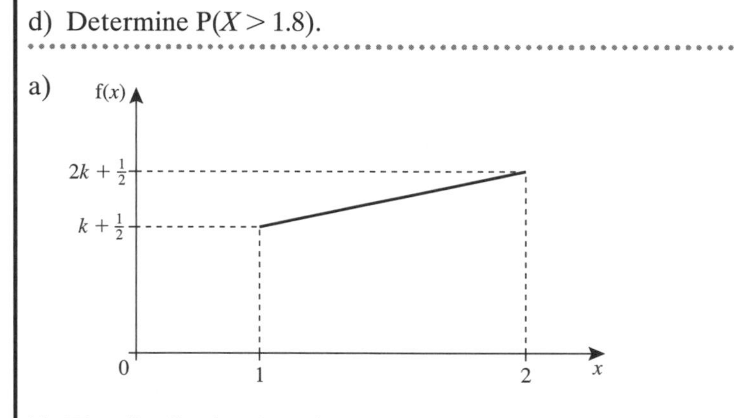

b) The distribution function takes the value 0 for $x \leqslant 1$ and the value 1 for $x \geqslant 2$. For $1 \leqslant x \leqslant 2$, F($x$) is given by:

$$\begin{aligned} F(x) &= \int_1^x \left(kx + \frac{1}{2}\right) dx \\ &= \left[\frac{1}{2}kx^2 + \frac{1}{2}x\right]_1^x \\ &= \frac{1}{2}k(x^2 - 1^2) + \frac{1}{2}(x - 1) \end{aligned}$$

Thus:

$$F(x) = \begin{cases} 0 & x \leqslant 1 \\ \frac{1}{2}k(x^2 - 1) + \frac{1}{2}(x - 1) & 1 \leqslant x \leqslant 2 \\ 1 & x \geqslant 2 \end{cases}$$

c) Since F is continuous with F(2) = 1, k is the solution of:

$$\begin{aligned} 1 &= \frac{1}{2}k(2^2 - 1) + \frac{1}{2}(2 - 1) \\ &= \frac{3}{2}k + \frac{1}{2} \end{aligned}$$

Thus:

$$\frac{3}{2}k = 1 - \frac{1}{2} = \frac{1}{2}$$

giving the solution $k = \frac{1}{3}$.

d) Substituting $k = \frac{1}{3}$ into the expression for F(x), for $1 \leqslant x \leqslant 2$, gives:

$$F(x) = \frac{1}{6}(x^2 - 1) + \frac{1}{2}(x - 1)$$

Thus:

$$\begin{aligned} P(X > 1.8) &= 1 - F(1.8) \\ &= 1 - \left\{\frac{1}{6}(1.8^2 - 1) + \frac{1}{2}(1.8 - 1)\right\} \\ &= 1 - \left(\frac{1}{6} \times 2.24 + \frac{1}{2} \times 0.8\right) \\ &= 1 - \frac{2.24 + 2.4}{6} \\ &= 1 - 0.7733 = 0.2267 \end{aligned}$$

The probability that X exceeds 1.8 is 0.227 (to 3 dp).

Exercise 3C

In Questions 1–6, the continuous random variable X has probability density function f and k is a positive constant.

1 Given that:

$$f(x) = \begin{cases} kx^2 + \frac{1}{6} & 0 < x < 3 \\ 0 & \text{otherwise} \end{cases}$$

sketch the graph of f. Find:

a) the value of k b) $P(X < 1)$ c) $P(X > 2)$.

2 Given that

$$f(x) = \begin{cases} kx^2 & -2 < x < 2 \\ 0 & \text{otherwise} \end{cases}$$

sketch the graph of f.

Find:

a) $F(x)$ in terms of k b) the value of k

c) $P(X > 1)$ d) $P(|X| < 1)$

e) $P(-1 < X < 0)$.

3 Given that:

$$f(x) = \begin{cases} \frac{1}{2}x & 1 < x < k \\ 0 & \text{otherwise} \end{cases}$$

sketch the graph of f.

Find:

a) $F(x)$ b) the value of k c) $P\left(\frac{3}{2} < X < 2\right)$

d) $P(X > 3)$ e) $P(2 < X < 3)$.

4 Given that:

$$f(x) = \begin{cases} 1 - kx & 0 < x < 2 \\ 0 & \text{otherwise} \end{cases}$$

sketch the graph of f.

Find:

a) the value of k b) $P(X \leqslant 1)$.

5 Given that:

$$f(x) = \begin{cases} 10k & -0.05 < x < 0.05 \\ 0 & \text{otherwise} \end{cases}$$

sketch the graph of f.

Find:

a) the value of k b) $P(X > 0.1)$ c) $P(X < 0.025)$.

6 Given that:

$$f(x) = \begin{cases} kx(6 - x) & 2 < x < 5 \\ 0 & \text{otherwise} \end{cases}$$

sketch the graph of f.

Find:

a) the value of k b) the mode, x_m c) $P(X < x_m)$.

7 A garage is supplied with petrol once a week. Its volume of weekly sales, X, in thousands of gallons, is distributed with probability density function given by:

$$f(x) = \begin{cases} kx(1 - x)^2 & 0 < x < 1 \\ 0 & \text{otherwise} \end{cases}$$

a) Determine the value of the constant k.

b) Determine an expression for the probability that the sales are less than c hundred gallons.

c) Determine the value of this probability for each of $c = 7$, 7.5 and 8.

d) Hence, or otherwise, determine an appropriate capacity for the garage's tank, if the probability that it is exhausted in a randomly chosen week is to be about 0.05.

3.4 The median, quartiles and percentiles

The **median**, m, is the value that bisects a distribution so that X is equally likely to be smaller or larger than m. Hence:

$$\int_{-\infty}^{m} \mathrm{f}(x)\,\mathrm{d}x = \int_{m}^{\infty} \mathrm{f}(x)\,\mathrm{d}x = 0.5 \qquad (3.8)$$

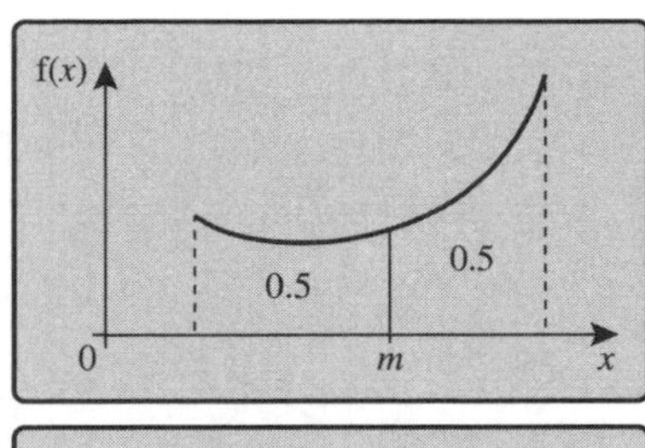

An equivalent definition is that m satisfies $\mathrm{F}(m) = 0.5$.

Example 9

The continuous random variable X has probability density function given by

$$\mathrm{f}(x) = \begin{cases} k(3-x) & 1 < x < 2 \\ k & 2 < x < 3 \\ k(x-2) & 3 < x < 4 \\ 0 & \text{otherwise} \end{cases}$$

where k is a positive constant.

a) Sketch the graph of f.

b) Determine the median m.

a)

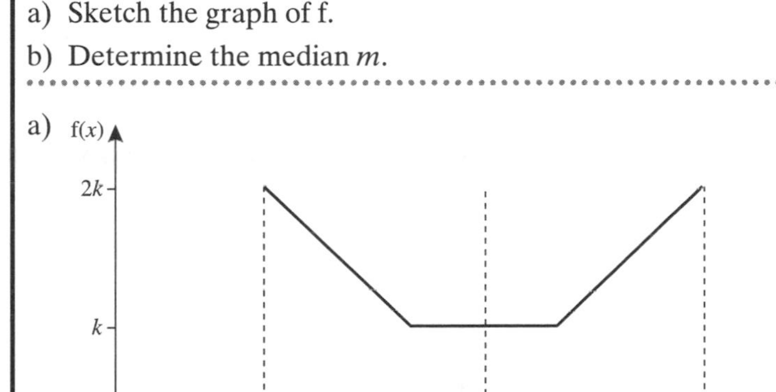

If the graph of f is symmetric about the line $x = x_0$, then the median is x_0.

b) The sketch shows that the probability density function is symmetrical about the line $x = 2.5$. The line of symmetry divides the area into two halves. Thus the median m is 2.5.

In this case, it is unnecessary to find the value of k.

Example 10

The continuous random variable X has probability density function given by:

$$\mathrm{f}(x) = \begin{cases} k - 5 + x & 5 < x < 6 \\ 0 & \text{otherwise} \end{cases}$$

where k is a positive constant.

a) Sketch the graph of f.

b) Determine the median.

a)

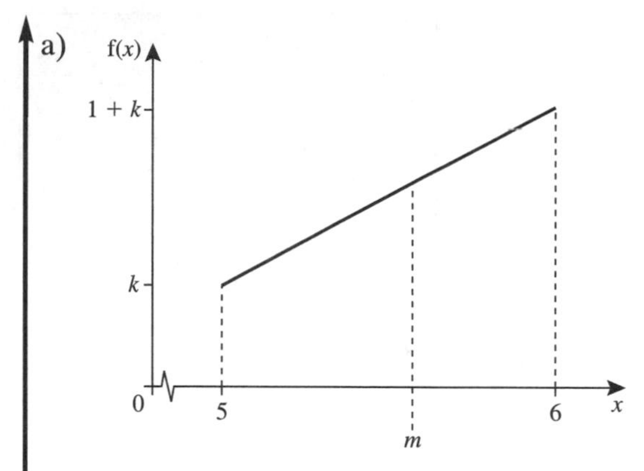

b) This time the probability density function is not symmetric and the most straightforward approach uses calculus. To determine the value of k, the fact that the total area is 1 is used. The area is given by:

$$\int_5^6 (k - 5 + x)\,\mathrm{d}x = \left[(k-5)x + \frac{1}{2}x^2\right]_5^6$$

$$= \left\{6(k-5) + \frac{1}{2} \times 6^2\right\} - \left\{5(k-5) + \frac{1}{2} \times 5^2\right\}$$

$$= (k-5) + 18 - \frac{25}{2}$$

$$= k + \frac{1}{2}$$

Since $\left(k + \frac{1}{2}\right) = 1$ it follows that $k = \frac{1}{2}$ and so:

$$\mathrm{f}(x) = x - \frac{9}{2} \quad \text{for } 5 < x < 6$$

To find the median, m, requires the solution of $\mathrm{F}(m) = \frac{1}{2}$:

$$\frac{1}{2} = \int_5^m \left(x - \frac{9}{2}\right)\mathrm{d}x$$

$$= \left[\frac{1}{2}x^2 - \frac{9}{2}x\right]_5^m$$

$$= \frac{1}{2}[x^2 - 9x]_5^m$$

$$= \frac{1}{2}\{(m^2 - 9m) - (25 - 9 \times 5)\}$$

$$= \frac{1}{2}(m^2 - 9m + 20)$$

Rearrangement gives:

$$m^2 - 9m + 19 = 0$$

with solution:

$$m = \frac{9 \pm \sqrt{81 - 76}}{2} = \frac{9 \pm \sqrt{5}}{2}$$

The root required lies between 5 and 6 (since this is the range of possible values for X) and so it is the larger root, $\frac{1}{2}(9 + \sqrt{5})$, that is relevant. To three significant figures, the median is 5.62.

Percentiles and **quartiles** are defined in a similar fashion to the median. For example, the 90th percentile is the solution of $F(x) = 0.90$, while the upper quartile is the solution of $F(x) = 0.75$.

Percentiles split a distribution into 100 equal parts; quartiles split a distribution into four equal parts.

S2

Example 11

The continuous random variable X has distribution function given by:

$$F(x) = \begin{cases} 0 & x \leqslant 0 \\ \frac{1}{4}x^2 & 0 \leqslant x \leqslant 2 \\ 1 & x \geqslant 2 \end{cases}$$

a) Determine l, the lower quartile of the distribution.

b) Determine d, the 95th percentile of the distribution.

a) The lower quartile is the solution of $F(l) = 0.25$. So:

$$\frac{1}{4}l^2 = 0.25 = \frac{1}{4}$$

Thus $l^2 = 1$. Since X only takes values between 0 and 2, the solution is $l = 1$ (and not $l = -1$).

The lower quartile is 1.

b) The 95th percentile is the solution of $F(d) = 0.95$. So:

$$\frac{1}{4}d^2 = 0.95$$

Thus $d^2 = 4 \times 0.95 = 3.8$. Hence d is the (positive) square root of 3.8, which is 1.95 (to 3 sf).

The 95th percentile is 1.95 (to 3 sf).

Example 12

The continuous random variable X has probability density function given by:

$$f(x) = \begin{cases} k & 2 < x < 3 \\ k(x - 2) & 3 < x < 4 \\ 0 & \text{otherwise} \end{cases}$$

where k is a positive constant.

a) Sketch the graph of f.

b) Determine the value of k.

c) Determine $P(X < 3)$.

d) Determine d, the 90th percentile.

a)

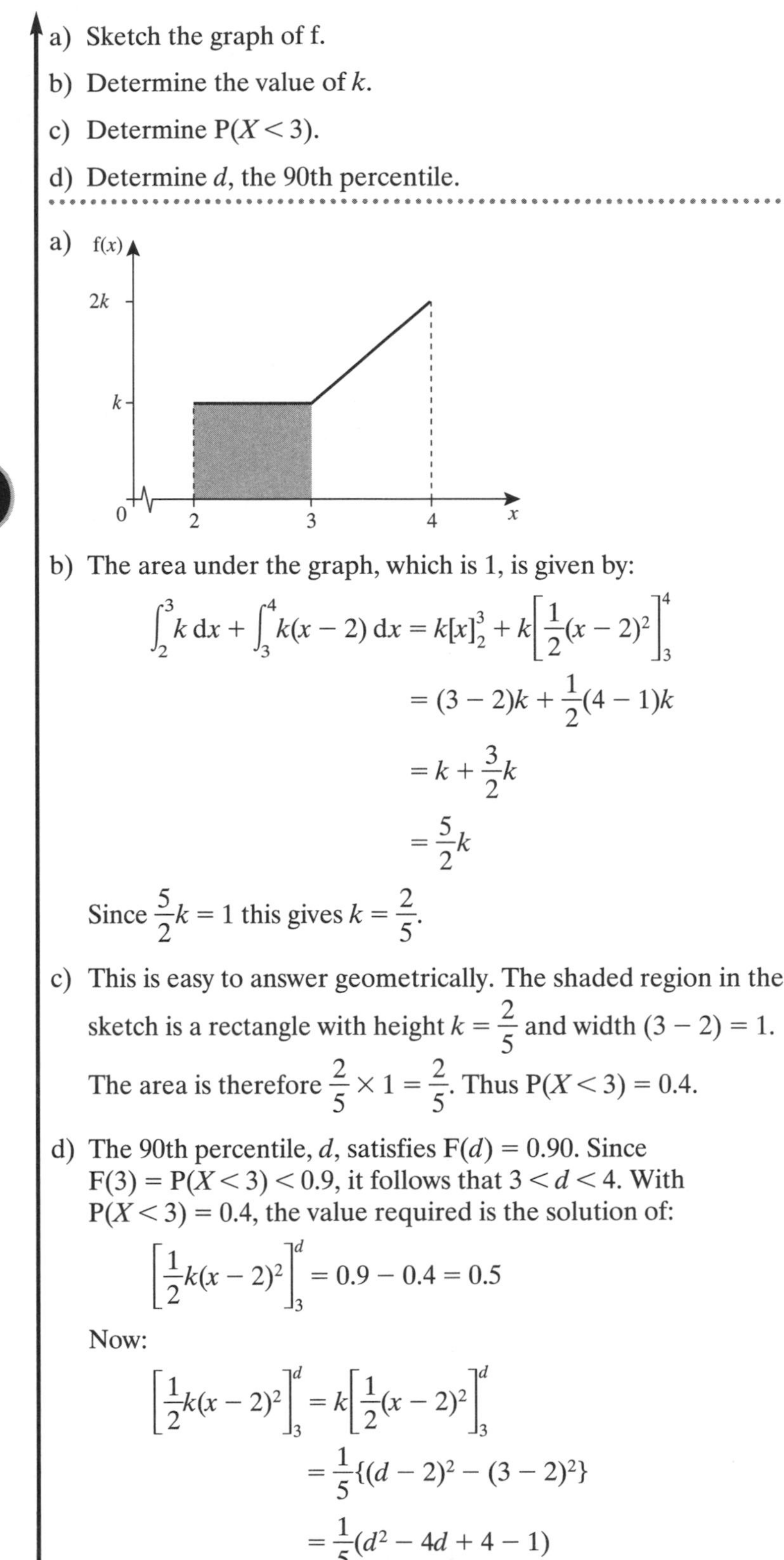

b) The area under the graph, which is 1, is given by:

$$\int_2^3 k\,\mathrm{d}x + \int_3^4 k(x-2)\,\mathrm{d}x = k[x]_2^3 + k\left[\frac{1}{2}(x-2)^2\right]_3^4$$

$$= (3-2)k + \frac{1}{2}(4-1)k$$

$$= k + \frac{3}{2}k$$

$$= \frac{5}{2}k$$

Since $\frac{5}{2}k = 1$ this gives $k = \frac{2}{5}$.

This could be answered geometrically.

c) This is easy to answer geometrically. The shaded region in the sketch is a rectangle with height $k = \frac{2}{5}$ and width $(3-2) = 1$. The area is therefore $\frac{2}{5} \times 1 = \frac{2}{5}$. Thus $P(X < 3) = 0.4$.

d) The 90th percentile, d, satisfies $F(d) = 0.90$. Since $F(3) = P(X < 3) < 0.9$, it follows that $3 < d < 4$. With $P(X < 3) = 0.4$, the value required is the solution of:

$$\left[\frac{1}{2}k(x-2)^2\right]_3^d = 0.9 - 0.4 = 0.5$$

Now:

$$\left[\frac{1}{2}k(x-2)^2\right]_3^d = k\left[\frac{1}{2}(x-2)^2\right]_3^d$$

$$= \frac{1}{5}\{(d-2)^2 - (3-2)^2\}$$

$$= \frac{1}{5}(d^2 - 4d + 4 - 1)$$

and hence d is the solution of:

$$\frac{1}{5}(d^2 - 4d + 3) = \frac{1}{2}$$

Cross-multiplying gives:

$$2(d^2 - 4d + 3) = 5$$

which simplifies to:

$$2d^2 - 8d + 1 = 0$$

and has solution:

$$d = \frac{-(-8) \pm \sqrt{8^2 - 4 \times 2}}{4} = 2 \pm \frac{1}{4}\sqrt{56} = 2 \pm \sqrt{3.5}$$

Since $d > 3$, the larger solution is required: the 90th percentile is 3.87 (to 3 sf).

S2

Example 13

The continuous random variable X has probability density function given by:

$$f(x) = \begin{cases} a(4+x) & -3 < x < -1 \\ -3ax & -1 < x < 0 \\ 0 & \text{otherwise} \end{cases}$$

where a is a positive constant.

a) Sketch the graph of f.

b) Determine the value of the constant a.

c) Determine the value of l, the lower quartile of the distribution.

a)

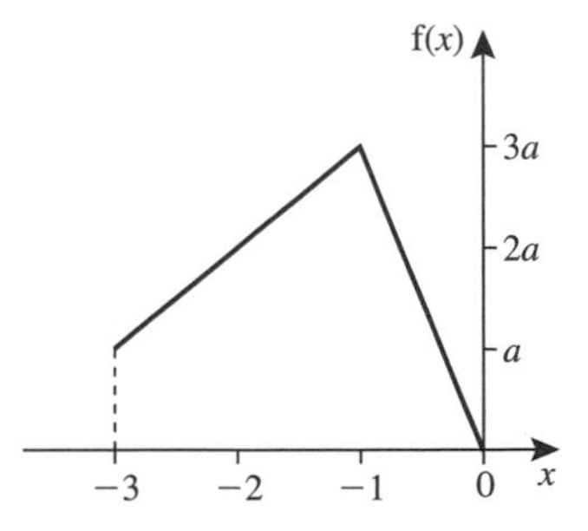

Despite the various minus signs, the probability density function is (as ever) never negative.

b) The value of a is found by using the result that the total area must equal 1. This total area is given by:

$$\int_{-3}^{-1} a(4+x)\,dx + \int_{-1}^{0} -3ax\,dx$$

$$= a\left[\frac{1}{2}(4+x)^2\right]_{-3}^{-1} - 3a\left[\frac{1}{2}x^2\right]_{-1}^{0}$$

$$= \frac{a}{2}[(4+x)^2]_{-3}^{-1} - \frac{3}{2}a[x^2]_{-1}^{0}$$

$$= \frac{a}{2}(9-1) - \frac{3}{2}(0-1)a$$
$$= 4a + \frac{3}{2}a$$
$$= \frac{11}{2}a$$

with the conclusion that $a = \frac{2}{11}$.

c) The lower quartile l is such that $F(l) = P(X \leqslant l) = \frac{1}{4}$. A glance at the sketch suggests that the lower quartile is less than -1, so that the relevant equation is:

$$\int_{-3}^{l} a(4+x)\,dx = \frac{1}{4}$$

Now:

$$\int_{-3}^{l} a(4+x)\,dx = a\left[\frac{1}{2}(4+x)^2\right]_{-3}^{l}$$
$$= \frac{2}{11} \times \frac{1}{2}\{(4+l)^2 - 1\}$$
$$= \frac{1}{11}\{(4+l)^2 - 1\}$$

Thus l is the solution of:

$$\frac{1}{11}\{(4+l)^2 - 1\} = \frac{1}{4}$$

Multiplying through by 11 and rearranging gives:

$$(4+l)^2 = \frac{11}{4} + 1 = 3.75$$

The two possible solutions are $(4+l) = \sqrt{3.75}$ and $(4+l) = -\sqrt{3.75}$, giving the possible solutions for l as -2.06 and -5.94 (to 3 sf). Considering the graph of f it is apparent that the first of these is the correct solution: the lower quartile is -2.06 (to 3 sf).

The area under the graph consists of a trapezium (for $-3 < x < -1$) and a triangle (for $-1 < x < 0$).
The trapezium has area $\frac{1}{2} \times 2 \times (a + 3a) = 4a$.
The triangle has area $\frac{1}{2} \times 1 \times 3a = \frac{3}{2}a$.
Thus the total area ($=1$) is $4a + \frac{3}{2}a = \frac{11}{2}a$. Hence $a = \frac{2}{11}$.

S2

Exercise 3D

1 The continuous random variable X has probability density function given by:

$$f(x) = \begin{cases} \frac{1}{4}x & 1 < x < 3 \\ 0 & \text{otherwise} \end{cases}$$

a) Sketch the graph of f.

b) Determine the median, and the lower and upper quartiles, of X.

2 The continuous random variable Y has distribution function given by:

$$F(y) = \begin{cases} 0 & y \leqslant 0 \\ \frac{1}{2}y & 0 \leqslant y \leqslant 1 \\ \frac{1}{8}(y+3) & 1 \leqslant y \leqslant 5 \\ 1 & y \geqslant 5 \end{cases}$$

a) Sketch the graph of F.

b) Determine the median, and the lower and upper quartiles, of Y.

In Questions **3–8**, the probability density function and distribution function of the continuous random variable X are f and F respectively, and k is a positive constant.

3 Given that:

$$f(x) = \begin{cases} kx + \frac{1}{4} & 1 < x < 3 \\ 0 & \text{otherwise} \end{cases}$$

sketch the graph of f.

Find:

a) the value of k

b) $F(x)$

c) $P(X > 2)$

d) the median of X

e) the lower and upper quartiles of X.

4 Given that:

$$f(x) = \begin{cases} -kx & -2 < x < 0 \\ kx & 0 < x < 2 \\ 0 & \text{otherwise} \end{cases}$$

sketch the graph of f.

Find:

a) the value of k

b) the median of X

c) $F(x)$

d) $P(0 < X < 1)$.

5 Given that:

$$f(x) = \begin{cases} kx^2 & -2 < x < 1 \\ 0 & \text{otherwise} \end{cases}$$

sketch the graph of f.

Find:

a) the value of k

b) $F(x)$

c) the median of X.

6 It is given that:

$$F(x) = \begin{cases} 0 & x \leqslant 0 \\ \frac{k}{3}\{(x+2)^3 - 8\} & 0 \leqslant x \leqslant 1 \\ 1 & x \geqslant 1 \end{cases}$$

Find:

a) the value of k

b) the median of X

c) the 30th and 70th percentiles of X

d) $f(x)$.

7 Given that:

$$f(x) = \begin{cases} 2x + k & 3 < x < 4 \\ 0 & \text{otherwise} \end{cases}$$

sketch the graph of f.

Find:

a) the value of k

b) $F(x)$

c) the lower and upper quartiles of X.

8 Given that:

$$f(x) = \begin{cases} 2(1 - x) & 0 < x < k \\ 0 & \text{otherwise} \end{cases}$$

sketch the graph of f.

Find:

a) the value of k

b) $F(x)$

c) the median of X

d) the 20th and 80th percentiles of X.

9 At Wetville, the proportion of the sky covered in cloud, S, has probability density function:

$$f(s) = \begin{cases} k(3 + s) & 0 < s < 1 \\ 0 & \text{otherwise} \end{cases}$$

Sketch the graph of f.

Find:

a) the value of the constant k

b) $F(s)$

c) $P(S > 0.5)$

d) the median of S.

10 The time, T hours, required to erect a type of wooden garden shed has probability density function:

$$f(t) = \begin{cases} kt^2 & 5 < t < 8 \\ 0 & \text{otherwise} \end{cases}$$

Sketch the graph of f.

Find:

a) the value of the constant k

b) $F(t)$

c) the 5th and 95th percentiles of T

d) the probability that it takes between 6 hours and 7 hours 20 minutes to erect a shed.

11 The continuous random variable Z has probability density function:

$$f(z) = \begin{cases} a + bz & 0 < z < 1 \\ 0 & \text{otherwise} \end{cases}$$

a) Given that $F(0.5) = 0.6$, determine the values of the constants a and b.

Determine also:

b) the median of Z

c) the lower and upper quartiles of Z.

S2

3.5 Mean, variance and standard deviation

If, in a sample of size n, the value x occurs with frequency f, then the sample mean, $\bar{x}$, is given by:

$$\bar{x} = \frac{1}{n}\sum fx = \sum\left\{x\left(\frac{f}{n}\right)\right\}$$

where $\frac{f}{n}$ is the relative frequency of the value x.

For a *discrete* random variable, as the sample size increases, the relative frequency of the value x converges on the corresponding probability.

In the same way, for a *continuous* random variable, with f now denoting the frequency of occurrences for some range of values of x, the relative frequency again converges on the corresponding probability.

Using the probability density function, the probability that X takes a value in a small range (width δx) in the neighbourhood of x is given by:

$$P\left[\left(x - \frac{\delta x}{2}\right) < X < \left(x + \frac{\delta x}{2}\right)\right] \approx f(x) \times \delta x$$

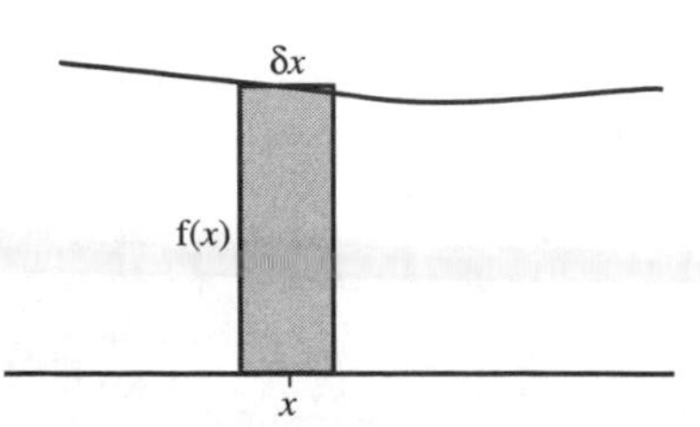

since the thin rectangle that approximates this probability has width δx and height $f(x)$.

The analogue of $\sum\left\{x\left(\frac{f}{n}\right)\right\}$ is therefore $\sum\{x(\mathrm{f}(x)\delta x)\}$, where the latter summation is over a huge number of values of x, each separated by a small amount δx. As the size of δx is decreased, so the value of $\sum\{x(\mathrm{f}(x)\delta x)\}$ converges on $\int x\mathrm{f}(x)\mathrm{d}x$. Hence, a continuous variable, X, has mean μ given by:

$$\mu = \int_{-\infty}^{\infty} x\mathrm{f}(x)\,\mathrm{d}x$$

The limits of the integral are given as $-\infty$ and ∞, but are in effect the largest and smallest possible values of X.

As in the case of a discrete random variable, μ is often called the expectation of X, denoted by $\mathrm{E}(X)$:

S2

$$\mathrm{E}(X) = \int_{-\infty}^{\infty} x\mathrm{f}(x)\,\mathrm{d}x \qquad (3.9)$$

If $\mathrm{f}(x)$ is symmetric about the line $x = c$ then $\mathrm{E}(X) = c$.

Expressions for the expectation of a function of X can be deduced by a corresponding argument to that used for $\mathrm{E}(X)$. Thus, for any general function g:

$$\mathrm{E}(\mathrm{g}(X)) = \int_{-\infty}^{\infty} \mathrm{g}(x)\mathrm{f}(x)\,\mathrm{d}x \qquad (3.10)$$

In particular:

$$\mathrm{E}(X^2) = \int_{-\infty}^{\infty} x^2\,\mathrm{f}(x)\,\mathrm{d}x \qquad (3.11)$$

Thus, since:

$$\mathrm{Var}(X) = \mathrm{E}((X-\mu)^2)$$

it can be calculated for a continuous random variable from:

$$\mathrm{Var}(X) = \int_{-\infty}^{\infty} (x-\mu)^2\,\mathrm{f}(x)\,\mathrm{d}x \qquad (3.12)$$

though it is usually easier to use:

$$\mathrm{Var}(X) = \mathrm{E}(X^2) - (\mathrm{E}(X))^2 = \int_{-\infty}^{\infty} x^2\,\mathrm{f}(x)\mathrm{d}x - \mu^2 \qquad (3.13)$$

The standard deviation of X is (as usual) the square root of $\mathrm{Var}(X)$.

All the results of Chapter 1 continue to hold. Thus, for constants a and b:

$$\mathrm{E}(aX+b) = a\mathrm{E}(X) + b \qquad (3.14)$$

$$\mathrm{Var}(aX+b) = a^2\,\mathrm{Var}(X) \qquad (3.15)$$

You should learn these formulae.

with the useful special cases:

$$\begin{aligned}
E(X+b) &= E(X)+b \\
E(aX) &= aE(X) \\
\mathrm{Var}(X+b) &= \mathrm{Var}(X) \\
\mathrm{Var}(aX) &= a^2\,\mathrm{Var}(X)
\end{aligned}$$

Example 14

The continuous random variable X has probability density function given by:

$$f(x) = \begin{cases} \frac{3}{4}(1-x^2) & -1 < x < 1 \\ 0 & \text{otherwise} \end{cases}$$

a) Sketch the graph of f and state the mode of X.

b) Determine $E(X)$ and $\mathrm{Var}(X)$.

c) Denoting these quantities by μ and σ^2 respectively, determine $P(\mu - \sigma < X < \mu + \sigma)$:

a)

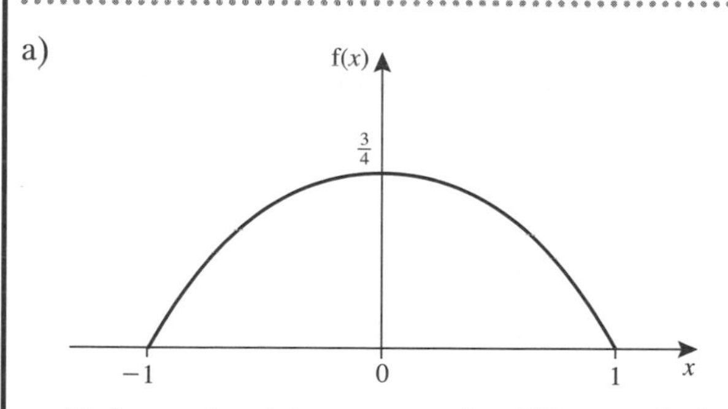

By inspection it is apparent that X has mode 0.

b) From the sketch it is apparent that $f(x)$ is symmetric about $x = 0$, hence $\mu = E(X) = 0$.

Calculation of $\mathrm{Var}(X)$ requires $E(X^2)$:

$$\begin{aligned}
E(X^2) &= \int_{-1}^{1} \frac{3}{4}x^2(1-x^2)\,dx \\
&= \frac{3}{4}\int_{-1}^{1}(x^2 - x^4)\,dx \\
&= \frac{3}{4}\left[\frac{1}{3}x^3 - \frac{1}{5}x^5\right]_{-1}^{1} \\
&= \frac{3}{4}\left\{\left(\frac{1}{3}-\frac{1}{5}\right) - \left(\frac{1}{3}\times(-1) - \frac{1}{5}\times(-1)\right)\right\} \\
&= \frac{3}{4}\left\{\left(\frac{1}{3}-\frac{1}{5}\right) - \left(-\frac{1}{3}+\frac{1}{5}\right)\right\} \\
&= \frac{3}{4}\left\{\left(\frac{1}{3}-\frac{1}{5}\right) + \left(\frac{1}{3}-\frac{1}{5}\right)\right\}
\end{aligned}$$

$$= \frac{3}{4} \times 2 \times \left(\frac{5-3}{3 \times 5}\right)$$

$$= \frac{3}{2} \times \frac{2}{15}$$

$$= \frac{1}{5}$$

Hence $\sigma^2 = \text{Var}(X) = \text{E}(X^2) - \mu^2 = \frac{1}{5}$.

c) Thus:

$$\text{P}(\mu - \sigma < X < \mu + \sigma) = \text{P}\left(-\sqrt{\frac{1}{5}} < X < \sqrt{\frac{1}{5}}\right)$$

$$= \frac{3}{4}\int_{-\sqrt{\frac{1}{5}}}^{\sqrt{\frac{1}{5}}} (1 - x^2)\,\text{d}x$$

$$= \frac{3}{2}\int_{0}^{\sqrt{\frac{1}{5}}} (1 - x^2)\text{d}x \text{ by symmetry}$$

The probability density function is symmetric about zero.

$$= \frac{3}{2}\left[x - \frac{1}{3}x^3\right]_0^{\sqrt{\frac{1}{5}}}$$

$$= \frac{3}{2}\left(\frac{1}{\sqrt{5}} - \frac{1}{3} \times \frac{1}{5\sqrt{5}}\right)$$

$$= \frac{3}{2} \times \frac{15 - 1}{15\sqrt{5}}$$

$$= \frac{3 \times 14}{2 \times 15\sqrt{5}}$$

$$= \frac{7}{5}\sqrt{\frac{1}{5}}$$

$$= 0.626 \text{ (to 3 dp)}$$

The required probability is 0.626, which is illustrated in the sketch.

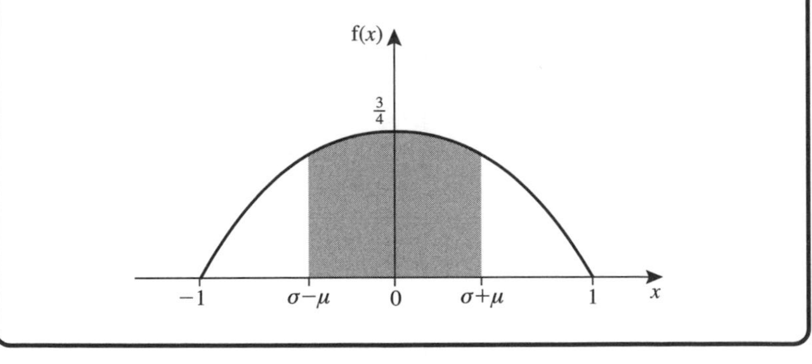

Example 15

The continuous random variable X has probability density function given by:

$$f(x) = \begin{cases} \frac{2}{3}(x+1) & 0 < x < 1 \\ 0 & \text{otherwise} \end{cases}$$

a) Sketch the graph of f.

b) Determine $E(X)$.

c) Determine $Var(X)$.

d) Determine the probability that two independent observed values of X both have values below the mean.

a)

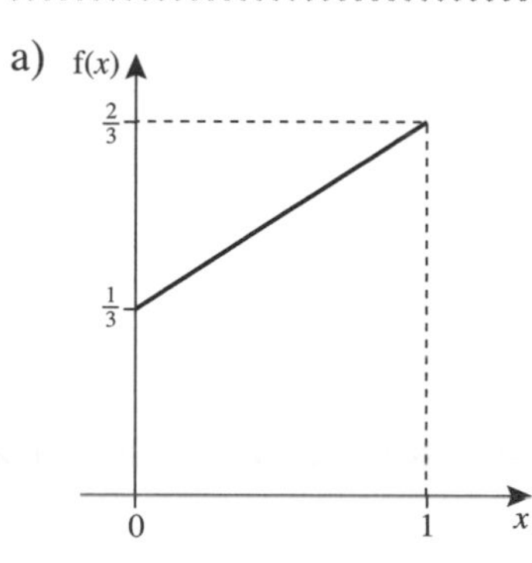

b) Since f(x) is not symmetric, integration is required:

$$\begin{aligned} E(X) &= \int_0^1 x \times \frac{2}{3}(x+1)\,dx \\ &= \frac{2}{3}\int_0^1 (x^2 + x)dx \\ &= \frac{2}{3}\left[\frac{1}{3}x^3 + \frac{1}{2}x^2\right]_0^1 \\ &= \frac{2}{3}\left(\frac{1}{3} + \frac{1}{2}\right) \\ &= \frac{2}{3} \times \frac{5}{6} \\ &= \frac{5}{9} \end{aligned}$$

c) In order to calculate the variance, $E(X^2)$ must be determined:

$$\begin{aligned} E(X^2) &= \int_0^1 x^2 \times \frac{2}{3}(x+1)\,dx \\ &= \frac{2}{3}\int_0^1 (x^3 + x^2)\,dx \\ &= \frac{2}{3}\left[\frac{1}{4}x^4 + \frac{1}{3}x^3\right]_0^1 \end{aligned}$$

S2

$$= \frac{2}{3}\left(\frac{1}{4} + \frac{1}{3}\right)$$
$$= \frac{2}{3} \times \frac{7}{12}$$
$$= \frac{7}{18}$$

Hence:

$$\text{Var}(X) = \text{E}(X^2) - (\text{E}(X))^2 = \frac{7}{18} - \left(\frac{5}{9}\right)^2 = \frac{63}{162} - \frac{50}{162}$$
$$= \frac{13}{162}$$

d) The probability that an observed value of X is less than the mean is given by:

S2

$$\text{P}\left(X < \frac{5}{9}\right) = \int_0^{\frac{5}{9}} \frac{2}{3}(x + 1)\,\text{d}x$$
$$= \frac{2}{3}\left[\frac{1}{2}x^2 + x\right]_0^{\frac{5}{9}}$$
$$= \frac{2}{3}\left(\frac{25}{162} + \frac{5}{9}\right)$$
$$= \frac{2}{3} \times \frac{115}{162}$$
$$= \frac{115}{243}$$

The probability that two independent observed values of X are both smaller than the mean is therefore (using the multiplication rule): $\left(\frac{115}{243}\right)^2 = 0.224$ (to 3 dp).

If events *A* and *B* are independent, then the probability that both occur is the product of their individual probabilities.

Example 16

The continuous random variable X has probability density function given by:

$$\text{f}(x) = \begin{cases} 4x^3 & 0 < x < 1 \\ 0 & \text{otherwise} \end{cases}$$

a) Sketch the graph of f.

b) Determine $\text{E}(X^{-1})$.

c) Determine $\text{Var}(X^{-1})$.

d) State the value of $\text{Var}\left(\frac{1 + X}{X}\right)$.

a)

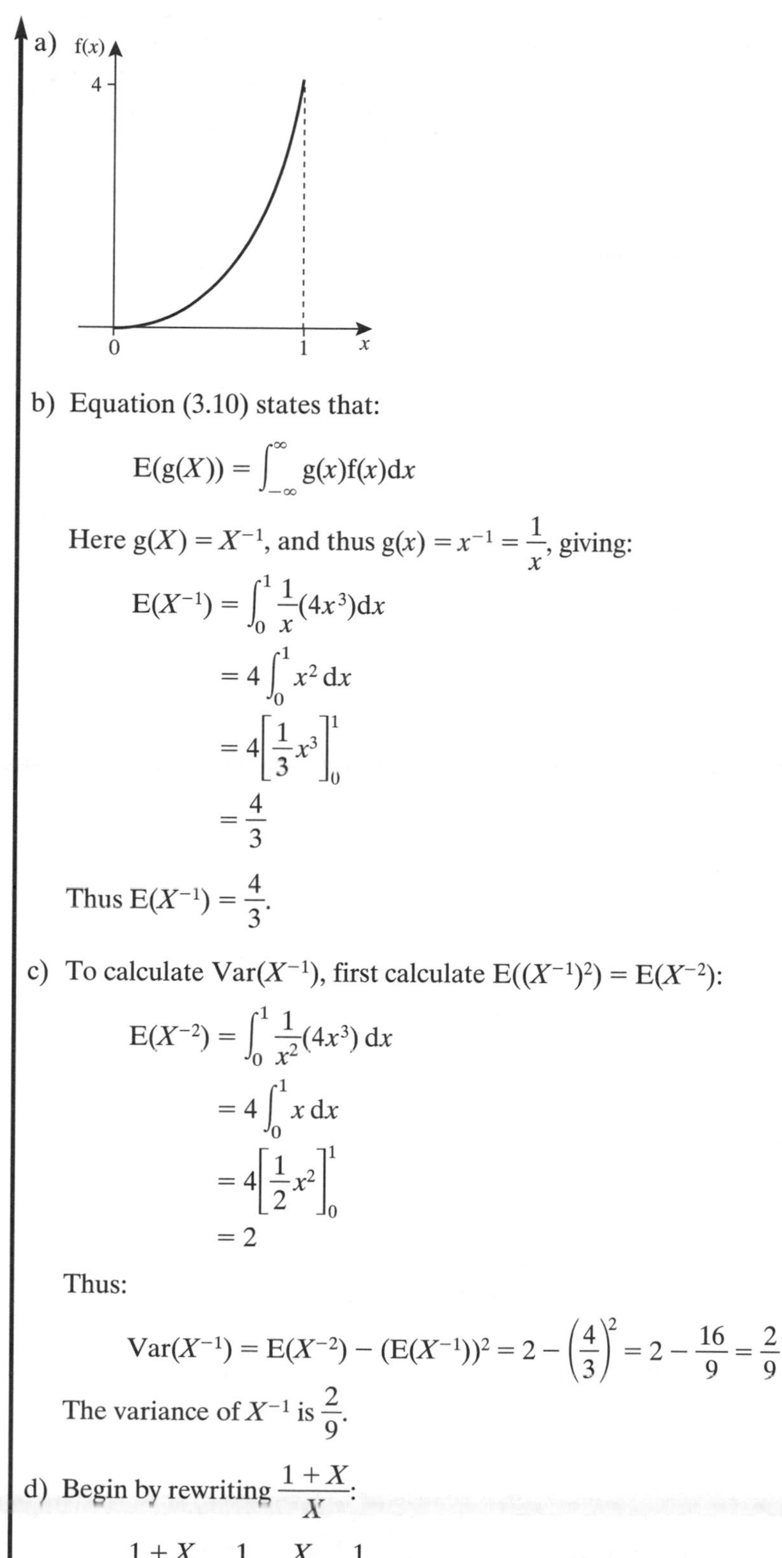

b) Equation (3.10) states that:

$$E(g(X)) = \int_{-\infty}^{\infty} g(x)f(x)dx$$

Here $g(X) = X^{-1}$, and thus $g(x) = x^{-1} = \frac{1}{x}$, giving:

$$E(X^{-1}) = \int_0^1 \frac{1}{x}(4x^3)dx$$
$$= 4\int_0^1 x^2\,dx$$
$$= 4\left[\frac{1}{3}x^3\right]_0^1$$
$$= \frac{4}{3}$$

Thus $E(X^{-1}) = \frac{4}{3}$.

c) To calculate $\text{Var}(X^{-1})$, first calculate $E((X^{-1})^2) = E(X^{-2})$:

$$E(X^{-2}) = \int_0^1 \frac{1}{x^2}(4x^3)\,dx$$
$$= 4\int_0^1 x\,dx$$
$$= 4\left[\frac{1}{2}x^2\right]_0^1$$
$$= 2$$

Thus:

$$\text{Var}(X^{-1}) = E(X^{-2}) - (E(X^{-1}))^2 = 2 - \left(\frac{4}{3}\right)^2 = 2 - \frac{16}{9} = \frac{2}{9}$$

The variance of X^{-1} is $\frac{2}{9}$.

d) Begin by rewriting $\frac{1+X}{X}$:

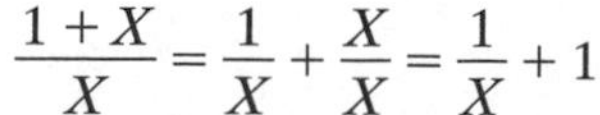

$$\frac{1+X}{X} = \frac{1}{X} + \frac{X}{X} = \frac{1}{X} + 1$$

S2

Since the addition of a constant has no effect on the variance:

$$\text{Var}\left(\frac{1+X}{X}\right) = \text{Var}\left(\frac{1}{X}+1\right) = \text{Var}\left(\frac{1}{X}\right) = \text{Var}(X^{-1}) = \frac{2}{9}$$

Thus $\text{Var}\left(\frac{1+X}{X}\right) = \frac{2}{9}$.

When a question requires you to 'State the value' you can expect that the answer follows almost immediately from previous working without the need for complicated further calculations.

Exercise 3E

1 The continuous random variable X has probability density function:

$$f(x) = \begin{cases} \frac{1}{2}x & 0 < x < 2 \\ 0 & \text{otherwise} \end{cases}$$

S2

Find:

a) $E(X)$

b) $E(X^2)$

c) $\text{Var}(X)$

d) $P(X < E(X))$.

2 The continuous random variable X has distribution function:

$$F(x) = \begin{cases} 0 & x \leqslant 1 \\ \frac{1}{8}(x^2 - 1) & 1 \leqslant x \leqslant 3 \\ 1 & x \geqslant 3 \end{cases}$$

Find:

a) $E(X)$

b) $\text{Var}(X)$.

3 The continuous random variable X has probability density function:

$$f(x) = \begin{cases} kx & 0 < x < 1 \\ k & 1 < x < 2 \\ k(3-x) & 2 < x < 3 \\ 0 & \text{otherwise} \end{cases}$$

Find:

a) the positive constant k

b) the mean, μ

c) the variance, σ^2

d) the median of X

e) $P(X < \mu)$.

f) Two independent observations of X are taken. Find the probability that one of the observations exceeds the mean and the other is less than the median.

4 The continuous random variable X has probability density function:

$$f(x) = \begin{cases} 2x & 0 < x < 1 \\ 0 & \text{otherwise} \end{cases}$$

Sketch the graph of f.

The random variable Y is defined by $Y = 4X + 2$. Find:

a) the expectation of Y

b) the standard deviation of Y.

5 The random variable X has probability density function:

$$f(x) = \begin{cases} x & 0 < x < 1 \\ k - x & 1 < x < 2 \\ 0 & \text{otherwise} \end{cases}$$

a) Find the value of the constant k.

b) Find the mean, μ, and show that the variance, σ^2, is $\frac{1}{6}$.

c) Determine the probability that a future observation takes a value between $(\mu - \sigma)$ and μ.

6 It is given that:

$$f(x) = \begin{cases} \frac{1}{4}(3x^2 - 6x + 4) & 0 < x < 2 \\ 0 & \text{otherwise} \end{cases}$$

Find:

a) $E(X)$ b) $Var(X)$.

7 The amount of cloud cover is measured on a scale from 0 to 1. A reasonable model for the amount of cloud cover, X, at mid-day during the spring is provided by the probability density function:

$$f(x) = \begin{cases} 8(x - 0.5)^2 + c & 0 < x < 1 \\ 0 & \text{otherwise} \end{cases}$$

with c being a non-negative constant.

a) Sketch the graph of f.

b) Calculate the value of c.

c) State the mean cloud cover.

S2

3.6 The rectangular distribution

A random variable having a **rectangular distribution** (also called a continuous uniform distribution) occurred in Example 1 of this chapter. For such a random variable, X, f(x) is constant for the entire range of possible values (from a to b, say):

$$f(x) = \begin{cases} \frac{1}{b-a} & a < x < b \\ 0 & \text{otherwise} \end{cases} \quad (3.16)$$

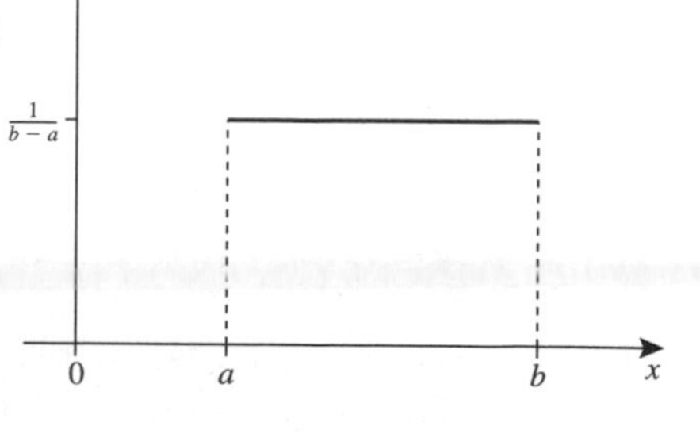

Between a and b the probability density is constant and the resulting shape of the graph is **rectangular**. The total probability of 1 is uniformly distributed over the interval $a \leq x \leq b$. The rectangle has width $(b - a)$ and height $\frac{1}{b-a}$, so that its area is equal to 1, as required.

Since f(x) is symmetrical about $x = \frac{1}{2}(a + b)$, the mean, E(X), and the median, m, are both equal to $\frac{1}{2}(a + b)$.

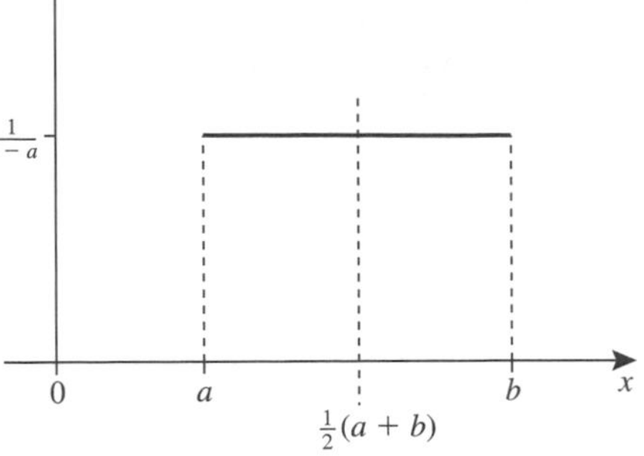

The distribution function, F, is given by:

$$\begin{aligned} F(x) = P(X \leqslant x) &= \int_a^x \frac{1}{b-a}\,dt \\ &= \left[\frac{t}{b-a}\right]_a^x \\ &= \frac{x-a}{b-a} \end{aligned}$$

Thus:

$$F(x) = \begin{cases} 0 & x \leqslant a \\ \dfrac{x-a}{b-a} & a \leqslant x \leqslant b \\ 1 & x \geqslant b \end{cases} \qquad (3.17)$$

To find the variance of X, the expectation of X^2 is required:

$$\begin{aligned} E(X^2) &= \int_a^b \frac{1}{b-a} x^2\,dx \\ &= \frac{1}{b-a}\left[\frac{1}{3}x^3\right]_a^b \\ &= \frac{b^3 - a^3}{3(b-a)} \\ &= \frac{1}{3}(b^2 + ab + a^2) \end{aligned}$$

$b^3 - a^3 = (b - a)(b^2 + ab + a^2)$

Since $E(X) = \frac{1}{2}(a + b)$:

$$\begin{aligned} \text{Var}(X) &= \frac{1}{3}(b^2 + ab + a^2) - \frac{1}{4}(a+b)^2 \\ &= \frac{1}{12}\{4(b^2 + ab + a^2) - 3(a^2 + 2ab + b^2)\} \\ &= \frac{1}{12}\{(4-3)b^2 + (4-6)ab + (4-3)a^2\} \\ &= \frac{1}{12}(b^2 - 2ab + a^2) \\ &= \frac{1}{12}(b-a)^2 \end{aligned}$$

A random variable, X, having a rectangular distribution over the interval ($a < x < b$) has:

$$E(X) = \frac{1}{2}(a + b) \qquad \text{Var}(X) = \frac{1}{12}(b-a)^2 \qquad (3.18)$$

Example 17

The distance between two points A and B is to be measured correct to the nearest tenth of a kilometre.

Determine the mean and standard deviation of the associated round-off error.

The rectangular distribution can be used to model rounding errors in measurements.

Suppose the length of AB is given as 45.2 km. The true length of AB could be any value between 45.15 km and 45.25 km. The round-off error (in km), X, could therefore take any value between -0.05 and 0.05. The random variable X therefore has a rectangular distribution with $b = 0.05$ and $a = -0.05$. The mean of the distribution is:

$$\frac{1}{2}(a+b) = \frac{1}{2}\{(-0.05) + 0.05\} = 0$$

and the variance of the distribution is

$$\frac{1}{12}(b-a)^2 = \frac{1}{12}\{0.05 - (-0.05)\}^2 = \frac{0.1^2}{12}$$

Thus the mean and standard deviation of the associated round-off error are 0 and $\dfrac{0.1}{\sqrt{12}} = 0.029$ (to 3 dp), respectively.

S2

Exercise 3F

1 The continuous random variable X has a rectangular distribution on the interval $0 < x < 2$. Find:

a) the probability density function of X

b) the distribution function of X.

2 The continuous random variable U has a rectangular distribution on the interval $a < u < b$. Given that $E(U) = 4$ and $Var(U) = 3$, find:

a) a and b b) $P(U > 5)$ c) $P(U < 2)$.

3 Given that:

$$f(x) = \begin{cases} 1 & 0 < x < 1 \\ 0 & \text{otherwise} \end{cases}$$

a) determine the expected value of $2X + 3$

b) determine the variance of $2X + 3$.

4 The continuous random variable T has a rectangular distribution on the interval $0 < t < 100$.

a) Find $P(20 < T < 60)$.

b) Denoting the mean and standard deviation of T by μ and σ respectively, find $P(|T - \mu| < \sigma)$.

5 The continuous random variable S has a rectangular distribution on the interval $c < s < d$. Given that $P(S < 3) = \frac{1}{4}$ and $P(S < 7) = \frac{3}{4}$, find c and d.

6 The continuous random variable Y has a rectangular distribution on the interval $0 < y < 2$ and $X = 3Y + 4$. Find:

a) $E(X)$ b) $Var(X)$.

7 Given that:

$$f(x) = \begin{cases} 4 & 0 < x < \frac{1}{4} \\ 0 & \text{otherwise} \end{cases}$$

sketch the graph of f.

Find:

a) $E(X)$ b) $E(2X - 4)$ c) $Var(X)$ d) $Var(2X - 4)$

8 Mrs Parent occasionally allows her daughter to borrow her car. When Mrs Parent leaves the car at home after driving it, the amount of petrol in the tank has a rectangular distribution between 10 litres and 50 litres. When her daughter leaves the car at home after having borrowed it, the amount of petrol in the tank has a rectangular distribution between 0 litres and 20 litres. (Mrs Parent is none too pleased!) Mrs Parent is the driver for 80% of the time and her daughter is the driver for the remaining 20% of the time. Mr Parent checks the car at home, not knowing who drove it last.
Find the probability that there is less than 15 litres of petrol in the tank.

S2

Summary

You should now be able to ...	Check out
1 Given a probability density function, determine the distribution function (and vice-versa).	**1** a) Determine the distribution function of X, given that $f(x) = \begin{cases} \frac{3}{16}(x-2)^2 & 0 < x < 4 \\ 0 & \text{otherwise} \end{cases}$ b) Determine the probability density function of X, given that: $F(x) = \begin{cases} 0 & x \leqslant -1 \\ \frac{1}{16}(x+1)^2 & -1 \leqslant x \leqslant 3 \\ 1 & x \geqslant 3 \end{cases}$

2 Determine the probability of an observation lying in a specified interval.	**2** Determine $P(1 < X < 2)$ for the cases in question **1**.
3 Determine the median, quartiles and percentiles of a continuous random variable.	**3** Determine: a) the median of X, where $f(x) = \begin{cases} \frac{1}{64}x^3 & 0 < x < 4 \\ 0 & \text{otherwise} \end{cases}$ b) the lower and upper quartiles, where $f(x) = \begin{cases} \frac{3}{2}(x-1)^2 & 0 < x < 2 \\ 0 & \text{otherwise} \end{cases}$ c) the ninth percentile of X, where $F(x) = \begin{cases} 0 & x \leqslant 0 \\ 0.1x & 0 \leqslant x \leqslant 10 \\ 1 & x \geqslant 10 \end{cases}$
4 Determine the mean, variance and standard deviation of a continuous random variable.	**4** Determine the mean, variance and standard deviation of the random variable X having probability density function: $f(x) = \begin{cases} \frac{1}{6}(6-x) & 2 < x < 4 \\ 0 & \text{otherwise} \end{cases}$
5 Calculate the mean, variance and standard deviation of a simple function of a continuous random variable	**5** a) Given that $E(X) = 4$, and that $E(X^2) = 25$, find the variance and standard deviation of X. b) Given that $E(X) = 12$ and $Var(X) = 16$, find: i) $E(X - 12)$ ii) $Var(X - 16)$ iii) $E(X^2)$ iv) $Var(-X)$. c) Given that $f(x) = \begin{cases} 3x^2 & 0 < x < 1 \\ 0 & \text{otherwise} \end{cases}$ find: i) $E(X^{-1})$ ii) $E(X^{-2})$ iii) $Var(X^{-1})$ iv) $E(10X^{-1})$ v) $Var(4X^{-1})$.
6 Recognise, and work with, a rectangular distribution	**6** The length of a wooden offcut is uniformly distributed between 0 and 5 cm. Determine the mean and variance of the length of an offcut.

S2

Revision exercise 3

1 The probability density function, $f(x)$, for the weight, X grams, of a banana is shown by the graph.

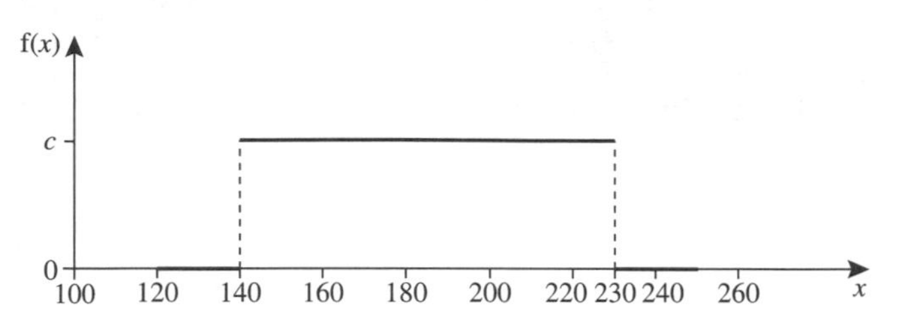

a) Find the value of the constant c, as used on the vertical axis.

b) Determine $P(X < 200)$.

c) Find values for the mean and variance of X.

d) Specify, giving a reason, an appropriate distribution for $\overline{X}$, the mean of a random sample of 75 observations of X. *(AQA, 2004)*

2 A town's library offers a booking system for free internet access. Users are allowed a maximum of one hour of connection time at any one booking.

The actual connection times, X minutes, of users may be modelled by the following probability density function, where k is a constant.

$$f(x) = \begin{cases} \dfrac{kx}{40} & 0 \leqslant x \leqslant 40 \\ k & 40 \leqslant x \leqslant 60 \\ 0 & \text{otherwise} \end{cases}$$

a) Sketch the graph of f.

b) By considering your sketch, or otherwise, show that the value of k is 0.025.

c) Hence determine $P(X > 30)$. *(AQA, 2004)*

3 The continuous random variable X has a rectangular distribution on the interval $(a < x < b)$, where $0 < a < b$.

a) Given that the mean, μ, is equal to 21 and that the variance, σ^2, is equal to 27, prove that $a = 12$ and $b = 30$.

b) Hence determine:

i) $P(5 < X < 20)$;

ii) $P\left(X < \mu - \dfrac{\sigma\sqrt{3}}{2}\right)$. *(AQA, 2004)*

4 At a hairdresser's, the time, T minutes, required to attend to a male customer may be modelled by the probability density function, $f(t)$, as represented by the graph.

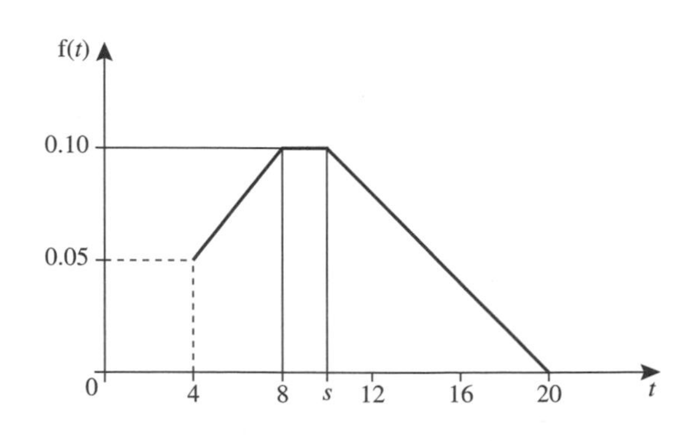

a) Determine $P(T < 8)$.

b) i) Show why the value of s, marked on the graph above, must be 10.

ii) Hence determine $P(T > 15)$.

(AQA, 2004)

5 The continuous random variable X has a rectangular distribution over the interval a to $a + k$, where a and k are positive constants.

a) Find, in terms of a and k, expressions for the mean and variance of X.

b) Given that $E(X) = 21$ and $Var(X) = 3$, show that $k = 6$ and $a = 18$.

c) Hence calculate $P(20 < X < 25)$. *(AQA, 2003)*

6 A continuous random variable T has the probability density function $f(t)$ given by:

$$f(t) = \begin{cases} \frac{1}{18}t^2 & 0 \leqslant t \leqslant 3 \\ \frac{1}{4}(5 - t) & 3 \leqslant t \leqslant 5 \\ 0 & \text{otherwise} \end{cases}$$

a) Sketch the graph of f.

b) Determine the distribution function, $F(t)$, for $0 \leqslant t \leqslant 5$.

c) Hence, or otherwise, calculate $P(T < 4)$. *(AQA, 2004)*

7 The time, T hours, taken by any member of a group of friends to complete a run for charity can be modelled by the following probability density function.

$$f(t) = \begin{cases} \frac{1}{90}t^2 & 3 \leqslant t \leqslant 6 \\ 2 - \frac{4}{15}t & 6 \leqslant t \leqslant 7.5 \\ 0 & \text{otherwise} \end{cases}$$

a) i) Show that the probability that a member of the group selected at random takes at least 6 hours to complete the run is 0.3.

ii) Evaluate the median time taken to complete the run.

b) Calculate the mean time taken to complete the run. *(AQA, 2004)*

8 All consultations with an optician are by appointment. The time, T minutes, by which an appointment is delayed has the following probability density function.

$$f(t) = \begin{cases} \frac{1}{54}t^2 & 0 \leqslant t < 3 \\ \frac{1}{6} & 3 \leqslant t < 6 \\ \frac{1}{24}(10 - t) & 6 \leqslant t < 10 \\ 0 & \text{otherwise} \end{cases}$$

a) Sketch the graph of f.

b) Use your graph to find:

i) $P(3 \leqslant T < 6)$; ii) $P(T > 6)$; iii) the lower quartile.

c) What proportion of appointments is delayed by less than 2 minutes? *(AQA, 2003)*

9 The random variable, X, has probability density function

$$f(x) = \begin{cases} cx + d & 0 < x < 2 \\ 0 & \text{otherwise} \end{cases}$$ where c and d are constants

a) Show that $2c + 2d = 1$.

b) Find the median of X:
  i) when $c = 0$ and $d = 0.5$;
  ii) when $c = 0.5$ and $d = 0$.

c) The following pairs of values are suggested for c and d.
  Suggestion A: $c = 0.5$ and $d = 0.5$
  Suggestion B: $c = -0.5$ and $d = 1$
  Suggestion C: $c = 1.5$ and $d = -1$

  Only one of the suggestions provides a valid probability density function.

  For **each** suggestion state whether or not it is valid. If you state it is not valid, give a reason. *(AQA, 2004)*

4 Estimation

This chapter will show you how to

- ✦ Use a t-distribution
- ✦ Construct confidence intervals for the mean of a normal population with unknown variance

Before you start

You should know how to ...	Check in
1 Calculate unbiased estimates of a population mean and variance.	**1** A random sample of 11 observations is summarized by $\sum x = 154$, $\sum(x - \bar{x})^2 = 248$. Calculate unbiased estimates of the population mean and variance.
2 Determine the distribution of the mean of a sample from a normal population.	**2** The random variable X has a normal distribution with mean 13 and variance 48. A random sample of 15 observations is drawn from this distribution. State the distribution of the sample mean, $\bar{X}$, giving its mean and variance.
3 Apply the Central Limit Theorem.	**3** The random variable Y has a distribution with mean 3 and variance 25. A random sample of 50 observations is drawn from this distribution. State the approximate distribution of the sample mean, $\bar{Y}$, giving its mean and variance.

In this chapter you will see how to obtain confidence intervals for the mean of a normal distribution, when the variance is *unknown*. As a reminder, the chapter begins with the procedure for the case where the variance is known.

4.1 Confidence intervals for the mean of a normal distribution with known variance

A random sample of n observations is taken from a $N(\mu, \sigma^2)$ population. The random variables corresponding to these observations are $X_1, X_2, \ldots, X_n$, with the random variable corresponding to the sample mean being denoted by $\bar{X}$. The distribution of $\bar{X}$ is a normal distribution with mean μ and variance $\frac{\sigma^2}{n}$.

You met this result in Chapter 5 of Statistics 1.

The random variable Z, given by:

$$Z = \frac{\bar{X} - \mu}{\frac{\sigma}{\sqrt{n}}}$$

has a $N(0, 1)$ distribution.

Table 4 in the Appendices gives the percentage points for a standard normal distribution.

For this standard normal distribution:

$\mathrm{P}(Z > 1.96) = 0.025$ $\quad$ $\mathrm{P}(Z < -1.96) = 0.025$ $\quad$ $\mathrm{P}(-1.96 < Z < 1.96) = 0.95$

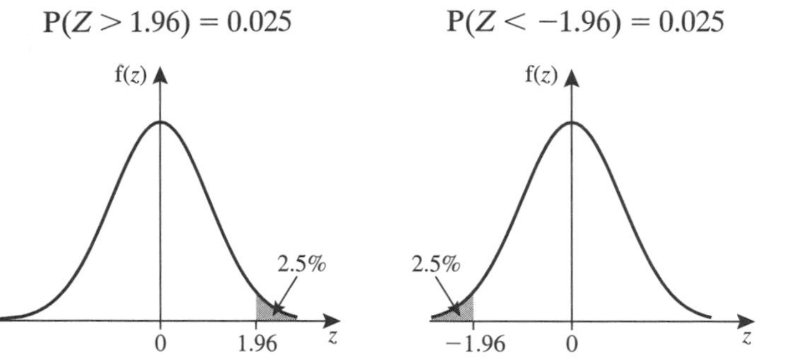

Substituting for Z, this implies that:

$$\mathrm{P}\left(-1.96 < \frac{\overline{X} - \mu}{\frac{\sigma}{\sqrt{n}}} < 1.96\right) = 0.95$$

Multiplying through the inequalities by $\frac{\sigma}{\sqrt{n}}$ gives:

$$\mathrm{P}\left(-1.96\frac{\sigma}{\sqrt{n}} < \overline{X} - \mu < 1.96\frac{\sigma}{\sqrt{n}}\right) = 0.95$$

This can be rewritten as:

$$\mathrm{P}\left(\overline{X} - 1.96\frac{\sigma}{\sqrt{n}} < \mu < \overline{X} + 1.96\frac{\sigma}{\sqrt{n}}\right) = 0.95$$

This is still a probability statement concerning the random variable $\overline{X}$. It is *not* a probability statement about the (unknown) constant μ.

Thus the probability that the interval:

$$\left(\overline{X} - 1.96\frac{\sigma}{\sqrt{n}}, \overline{X} + 1.96\frac{\sigma}{\sqrt{n}}\right)$$

contains the population mean is 95%.

Replacing $\overline{X}$ by $\bar{x}$, the sample mean, gives the interval:

Remember: $\overline{X}$ is the **random variable** corresponding to the sample mean; $\bar{x}$ is a particular value of the sample mean.

$$\left(\bar{x} - 1.96\frac{\sigma}{\sqrt{n}}, \bar{x} + 1.96\frac{\sigma}{n}\right) \qquad (4.1)$$

which is called a **95% symmetric confidence interval** for μ. The two limiting values that define the interval are known as the **confidence limits**.

Often the adjective 'symmetric' is omitted.

Different samples lead to different values of $\bar{x}$ and hence to different 95% confidence intervals. On average, 95% of these will include μ.

You met confidence intervals in Chapter 5 of Statistics S1.

The next diagram shows one hundred 95% confidence intervals, all based on samples of the same size and from the same population. Each has been calculated using Equation (4.1). The population mean value, μ, is indicated. It will be seen that most of the confidence intervals do overlap the true mean value. On average 95% will do so, though in this particular set of trials only 92 of the 100 intervals include μ. Those that do not do so are indicated by the bolder lines.

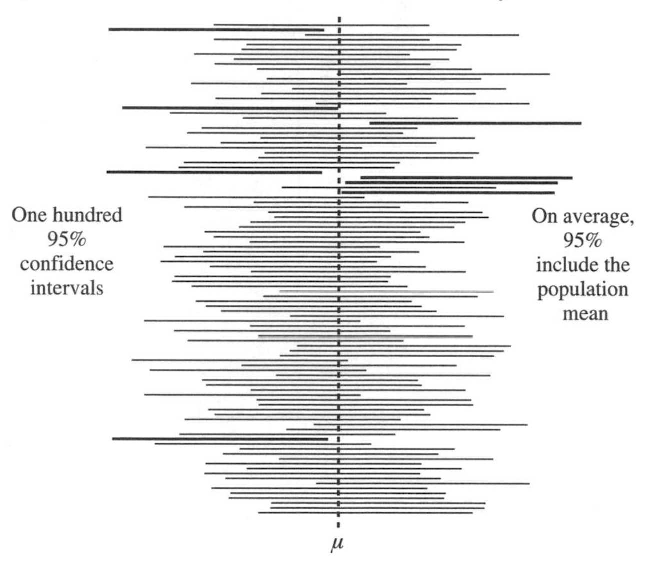

S2

4.2 The *t*-distribution

Suppose that the question of interest is, once again, the value of the unknown mean, μ, of a normal population. Suppose, however, that the population variance, σ^2, is also unknown. As usual, $X_1, X_2, \ldots, X_n$ denote the random variables corresponding to n observations randomly sampled from the population, with $\overline{X}$ corresponding to the sample mean.

If the variance had been known, then the confidence interval would have been based on the standard normal random variable Z given by:

$$Z = \frac{\overline{X} - \mu}{\frac{\sigma}{\sqrt{n}}}$$

See Section 4.1.

When σ is unknown, and n is large, the same procedure may be used, but with σ replaced by the sample standard deviation, s. However, although s^2 is an unbiased estimate of σ^2, its value varies from sample to sample. The random variable corresponding to s^2 is S^2. Replacement of σ by S in the previous expression for Z gives the random variable T defined by:

$$T = \frac{\overline{X} - \mu}{\frac{S}{\sqrt{n}}}$$

The formula for T therefore involves *two* random variables: $\overline{X}$ in the numerator and S in the denominator. The values of T will vary from sample to sample not only because of variations in $\overline{X}$ (as in the case of Z) but also because of variations in S.

The distribution of T is a ***t*-distribution**. A t-distribution is symmetric about zero and has a single parameter, ν, which is a positive integer and is known as the number of **degrees of freedom** of the distribution. In the case of T, $\nu = n - 1$.

The character ν comes from the Greek alphabet and is pronounced 'nu'.

As a shorthand the phrase 'a t-distribution with ν degrees of freedom' is written as 'a t_ν-distribution'. Thus T has a t_{n-1}-distribution.

The distribution of T has $n - 1$ rather than n degrees of freedom for the same reason that s^2, the unbiased estimate of σ^2, has divisor $(n - 1)$ rather than n. This occurs because s^2 involves $\Sigma(x - \bar{x})^2$, whereas σ^2 involves $\Sigma(x - \mu)^2$.

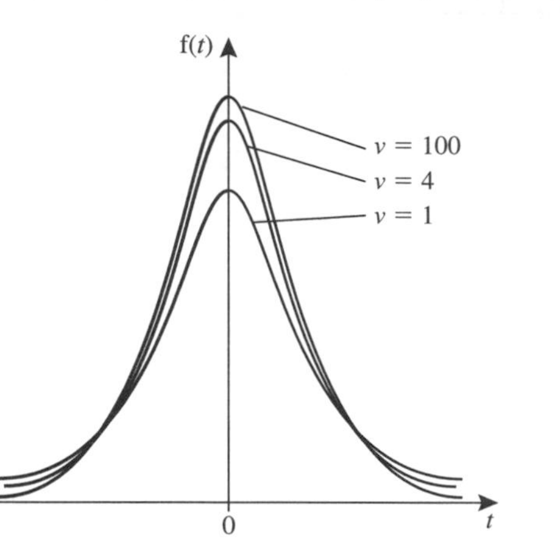

Thus, for a sample of size 5, the graph showing the probability density function with $\nu = 4$ is appropriate.

S2

As ν increases so the corresponding t_ν-distribution increasingly resembles the limiting standard normal distribution (which corresponds to $\nu = \infty$). When ν is 30 or more, the differences between the t_ν-distribution and the normal distribution are very slight.

The result 'T has a t_{n-1}-distribution' holds if $X_1, \ldots, X_n$ are independent random variables with the same normal distribution.

Example 1

The weights, in grams, of bags of wine gums are normally distributed with mean μ and variance σ^2. The trading inspector chooses a random sample of 16 bags for measurement in order to determine a confidence interval for the population mean. State a relevant random variable (in terms of the sample mean $\bar{X}$ and the unknown population mean, μ):

a) if the standard deviation is known to be 2,

b) if the standard deviation is unknown.

In each case you should state the relevant distribution.

The sample size, n, is 16.

a) If it is known that $\sigma = 2$, then an appropriate random variable is:

$$\frac{\bar{X} - \mu}{\frac{2}{\sqrt{16}}} = \frac{\bar{X} - \mu}{\frac{2}{4}} = 2(\bar{X} - \mu)$$

The relevant distribution is the standard normal distribution.

b) If σ^2 is unknown, the appropriate statistic is:

$$\frac{\overline{X} - \mu}{\sqrt{\frac{S^2}{16}}} = \frac{\overline{X} - \mu}{\frac{1}{4}S} = \frac{4(\overline{X} - \mu)}{S}$$

The relevant distribution is a t-distribution with 15 degrees of freedom.

Using the table of percentage points of the Student's t-distribution (Table 5)

Table 5 in the AQA tables gives the values of a random variable having a t_ν-distribution that correspond to certain cumulative probabilities. These values are called **percentage points** and are often referred to as **critical values**. They are given for many values of ν.

The man who first wrote about the t-distribution was William Gossett, in 1908. He published his findings using the pen-name 'Student'. The distribution is therefore often called the Student's t-distribution.

S2

Here is a brief extract from Table 5 (which is reproduced in the Appendices to this book):

p	**0.9**	**0.95**	**0.975**	**0.99**	**0.995**
ν					
1	3.078	6.314	12.706	31.821	63.657
2	1.886	2.920	4.303	6.965	9.925
3	1.638	2.353	3.182	4.541	5.841
.	.	.	.	.	.
.	.	.	.	.	.
.	.	.	.	.	.
28	1.313	1.701	2.048	2.467	2.763

P(T < 1.699) = 0.95 when ν = 29.

p	**0.9**	**0.95**	**0.975**	**0.99**	**0.995**
ν					
29	1.311	1.699	2.045	2.462	2.756
30	1.310	1.697	2.042	2.457	2.750
31	1.309	1.696	2.040	2.453	2.744
.	.	.	.	.	.
.	.	.	.	.	.
.	.	.	.	.	.
∞	1.282	1.645	1.960	2.326	2.576

This row corresponds to the normal distribution.

A problem with tables of the t-distribution is finding the correct column. You may find it helpful to begin by locating the corresponding normal critical value, which will be given (to 3 d.p.) in the final row of the table.

Example 2

The random variable T has a t-distribution with 3 degrees of freedom. Determine the values of t for which:

a) $P(T < t) = 0.99$ b) $P(T < t) = 0.10$

c) $P(|T| < t) = 0.90$ d) $P(|T| > t) = 0.05$.

a) The column headed '0.99' gives $t = 4.541$.

b) The column headed '0.9' gives $P(T < 1.638) = 0.90$, which implies that $P(T > 1.638) = 0.10$. By symmetry, $P(T < -1.638) = 0.10$, and hence the required value of t is -1.638.

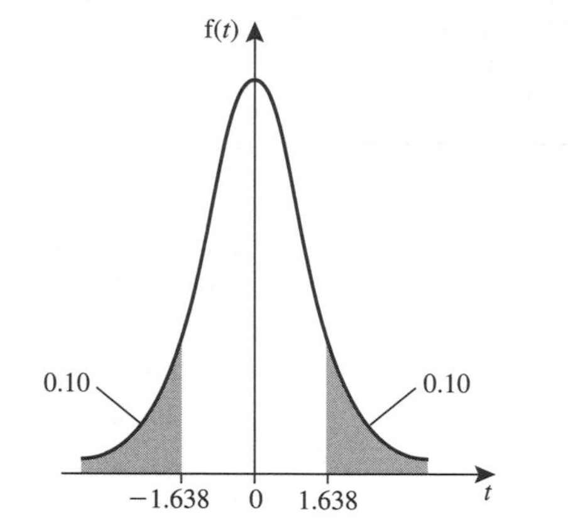

c) The required value for t is the value corresponding to an upper-tail probability of 0.05, and hence $p = 0.95$. The required value of t is therefore 2.353.

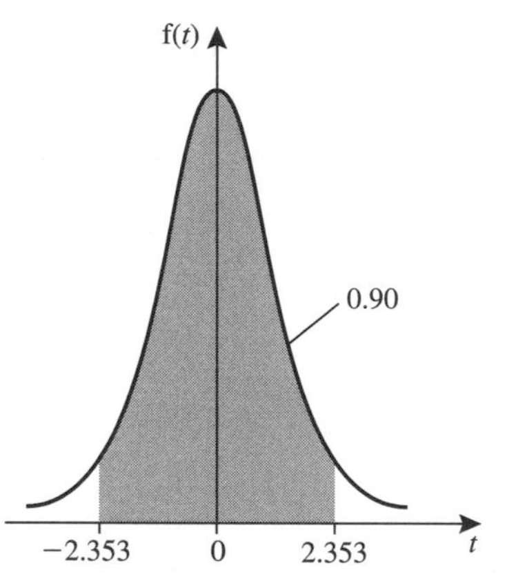

d) A combined probability of 0.05 in the two tails implies 0.025 in the upper tail, and hence $p = 0.975$. The value of t is 3.182.

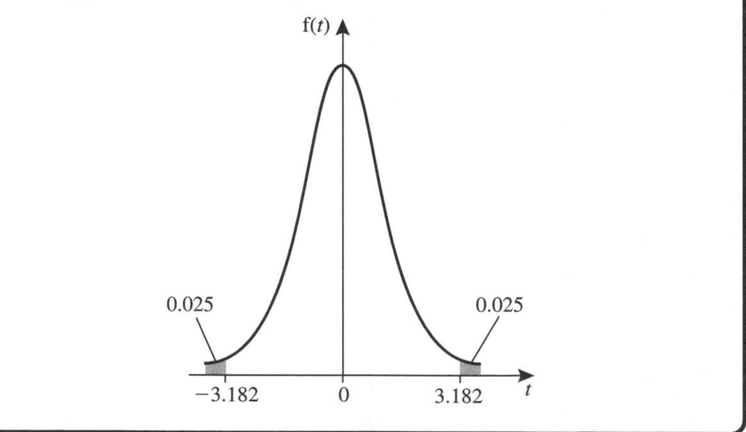

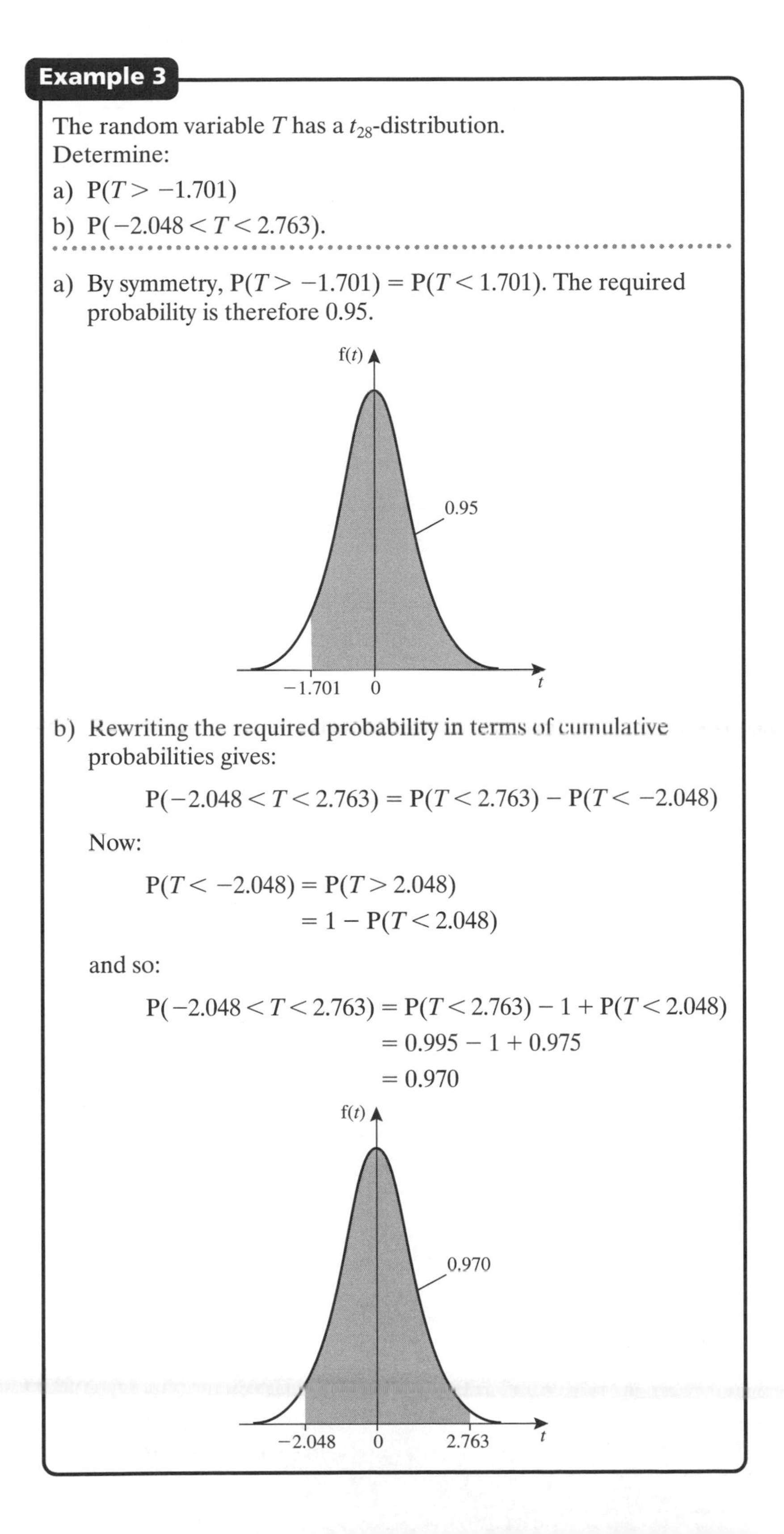

Example 3

The random variable T has a t_{28}-distribution.
Determine:

a) $P(T > -1.701)$

b) $P(-2.048 < T < 2.763)$.

a) By symmetry, $P(T > -1.701) = P(T < 1.701)$. The required probability is therefore 0.95.

b) Rewriting the required probability in terms of cumulative probabilities gives:

$$P(-2.048 < T < 2.763) = P(T < 2.763) - P(T < -2.048)$$

Now:

$$P(T < -2.048) = P(T > 2.048)$$
$$= 1 - P(T < 2.048)$$

and so:

$$P(-2.048 < T < 2.763) = P(T < 2.763) - 1 + P(T < 2.048)$$
$$= 0.995 - 1 + 0.975$$
$$= 0.970$$

Exercise 4A

1 The random variable T has a t-distribution with ν degrees of freedom. Find the value of t in each of the following cases:

a) $\nu = 8$, $P(T < t) = 0.95$

b) $\nu = 15$, $P(T > t) = 0.01$

c) $\nu = 5$, $P(|T| < t) = 0.95$

d) $\nu = 12$, $P(T < t) = 0.01$

e) $\nu = 4$, $P(T > t) = 0.975$

f) $\nu = 10$, $P(|T| > t) = 0.10$.

2 Given that T has a t_5-distribution, find:

a) $P(T < -2.764)$

b) $P(T > 1.812)$

c) $P(-2.228 < T < 1.372)$.

3 Given that T has a t_5-distribution, find:

a) $P(T < 2.571)$ b) $P(T > -4.032)$ c) $P(|T| < 2.015)$.

4.3 Confidence intervals for the mean of a normal distribution with unknown variance

A small sample has been taken from a normal distribution (population) with mean μ and with unknown variance. In this case, the random variable T, given by:

$$T = \frac{\overline{X} - \mu}{\frac{S}{\sqrt{n}}}$$

has a t_{n-1}-distribution.

Let c be the value such that:

$$P(T > c) = 0.025$$

The value of c is found from Table 5 of the Appendices.

Like the normal distribution, the t-distribution is symmetric about zero, so that:

$$P(T < -c) = 0.025$$

and thus:

$$P(-c < T < c) = 0.95$$

Substituting $\dfrac{\overline{X} - \mu}{\frac{S}{\sqrt{n}}}$ for T and rearranging the expression gives:

$$P\left(\overline{X} - c\frac{S}{\sqrt{n}} < \mu < \overline{X} + c\frac{S}{\sqrt{n}}\right) = 0.95$$

This is a statement about the random variables $\overline{X}$ and S. The mean μ is not a random variable.

The (symmetric) confidence interval for μ therefore becomes:

$$\left(\bar{x} - c\frac{s}{\sqrt{n}}, \bar{x} + c\frac{s}{\sqrt{n}}\right)$$

or, equivalently:

$$\left(\bar{x} - c\sqrt{\frac{s^2}{n}}, \bar{x} + c\sqrt{\frac{s^2}{n}}\right)$$

The adjective 'symmetric' is often omitted.

S2

Example 4

A random sample of 16 sweets are chosen from a box of sweets and the weight, x grams, of each sweet is determined. The measurements are summarised by $\sum x = 13.3$ and $\sum(x - \bar{x})^2 = 4.05$, where $\bar{x}$ is the sample mean.
Assuming that the weights have a normal distribution, determine a 99% confidence interval for the population mean, giving the limits to three decimal places.

Since the population has a normal distribution with unknown variance, it is appropriate to base a confidence interval on a t-distribution – in this case the t_{15}-distribution.

From Table 5 the critical value is found to be 2.947.

The sample mean is:

$$\bar{x} = \frac{13.3}{16} = 0.831\,25$$

and the unbiased estimate of the population variance, s^2, is:

$$\frac{1}{15} \times 4.05 = 0.27$$

The 99% confidence interval is therefore:

$$\left(0.831\,25 - 2.947 \times \sqrt{\frac{0.27}{16}}, 0.831\,25 + 2.947 \times \sqrt{\frac{0.27}{16}}\right)$$

which simplifies to:

(0.448, 1.214)

Thus (to 3 dp) the 99% confidence interval for the population mean is (0.448 grams, 1.214 grams).

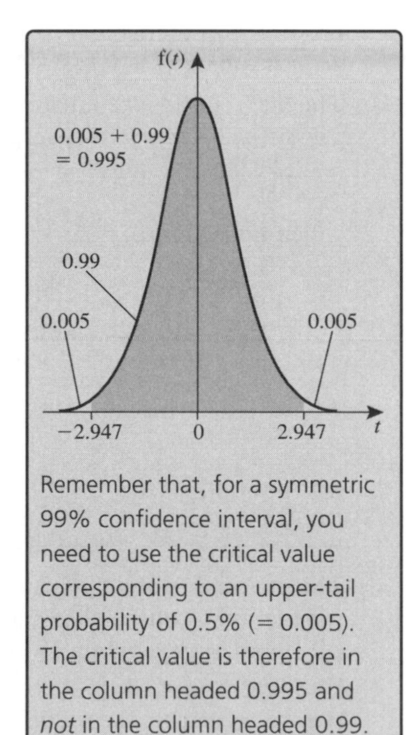

Remember that, for a symmetric 99% confidence interval, you need to use the critical value corresponding to an upper-tail probability of 0.5% (= 0.005). The critical value is therefore in the column headed 0.995 and *not* in the column headed 0.99.

Example 5

Ten students independently performed an experiment to estimate the value of π. Their results were

3.02, 3.06, 2.84, 3.23, 2.90, 3.01, 3.40, 2.71, 2.92, 3.00

a) Calculate the sample mean and standard deviation of these data.

b) Stating any assumption that you need to make, calculate a 90% confidence interval for π based on these data, giving the limits to two decimal places.

c) With the same assumption, calculate a 99% confidence interval for π, giving the limits to two decimal places.

d) Comment on your answers to parts b) and c).

a) A calculator gives the sample mean and standard deviation (s) as having the values 3.009 and 0.1950, respectively.

b) The assumption is that the population has a normal distribution with mean π.

For a symmetric 90% confidence interval, the critical value of t corresponds to a cumulative probability of 0.95. With 10 observations there are 9 degrees of freedom and hence the critical value is 1.833.

The 90% confidence interval becomes:

$$\left(3.009 - 1.833 \times \frac{0.1950}{\sqrt{10}},\ 3.009 + 1.833 \times \frac{0.1950}{\sqrt{10}}\right)$$

which simplifies to:

(2.90, 3.12) (to 2 dp)

c) For a 99% confidence interval, the critical value uses the 0.995 point of a t_9-distribution, which is 3.250. Thus the interval is:

$$\left(3.009 - 3.250 \times \frac{0.1950}{\sqrt{10}},\ 3.009 + 3.250 \times \frac{0.1950}{\sqrt{10}}\right)$$

which simplifies to:

(2.81, 3.21) (to 2 dp)

d) The probability of a 90% confidence interval including the population mean is 0.90, which implies that it fails to include the population mean on 10% of occasions.

If the method used by the students is unbiased (so that the population mean does equal 3.14 ...) then this occasion is one of the 10%. By contrast, the much wider 99% interval (which fails to include the mean on only 1% of occasions) does include 3.14.

Example 6

A recycling firm keeps records of the amounts of waste that are collected each week in a waste depository. Over a 39-week period it finds that the mean and standard deviation (s) of the weekly amount deposited are respectively 217 kg and 42 kg. The amounts deposited during these 39 weeks can be considered as being a random sample from the population of weekly amounts deposited.

a) Determine a 95% confidence interval for the population mean, assuming that the population standard deviation is 42 kg.

b) Determine a 95% confidence interval for the population mean, assuming that the population of weekly amounts is normal.

c) State, with a reason, why neither interval may provide a reliable guide to the annual amount of waste deposited.

a) As a consequence of the Central Limit Theorem, the sample mean will be an observation from a normal distribution with, according to the assumption, a variance of $\frac{42^2}{39}$. The 95% confidence interval for the mean is:

$$\left(217 - 1.960 \times \frac{42}{\sqrt{39}}, 217 + 1.960 \times \frac{42}{\sqrt{39}}\right) = (204, 230) \text{ to 3 sf}$$

Assuming that the population standard deviation is 42 kg, the 95% confidence interval for the mean amount of waste collected per week is from 204 kg to 230 kg (to 3 sf).

b) If the population is known to be normal, then, using s^2 as an unbiased estimate of σ^2, the t_{38}-distribution is relevant. From Table 5 of the Appendices, the f = 0.975 critical value is 2.024, so that the 95% confidence interval becomes:

$$\left(217 - 2.024 \times \frac{42}{\sqrt{39}}, 217 + 2.024 \times \frac{42}{\sqrt{39}}\right) = (203, 231) \text{ to 3 sf}$$

Assuming that the weekly amounts have a normal distribution, the 95% confidence interval for the mean amount of waste collected per week is from 203 kg to 231 kg (to 3 sf).

c) There are likely to be annual cycles in the amount of waste deposited: the 39-week period may be unrepresentative of the year as a whole.

A better scheme would have been to have sampled weekly for an entire year, or to have sampled 39 randomly chosen weeks.

S2

Calculator practice

Some calculators can calculate a confidence interval based on the t-distribution (and also on the normal distribution). The details will be given in the manual for the calculator. Typically you will need to switch to a special mode and select from the options available. You will be required to provide the sample data (either a set of sample values or the sample statistics $n, \bar{x}$ and s) and the confidence level.

Exercise 4B

1 The random variable X has a normal distribution with mean μ. A random sample of 10 observations of X is taken and gives $\sum x = 83.3$, $\sum(x - \bar{x})^2 = 27.52$, where $\bar{x}$ is the sample mean. Find:

a) a 95% confidence interval for μ

b) a 99% confidence interval for μ.

2 The quantity of milk in a carton may be assumed to have a normal distribution. A random sample of 16 cartons was taken and the quantity of milk was measured, with the following results, in ml.

1005, 1003, 998, 1001, 1002, 999, 1000, 1001, 1007, 1003,
1010, 1001, 1003, 1002, 1005, 995

a) Find a 99% confidence interval for the mean quantity of milk in a carton, giving the limits, in ml, to two decimal places.

b) Assuming that the values of the population mean and variance are equal to their unbiased estimates, by how much should the mean be increased if the law requires that no more than 10% of cartons can contain quantities less than 1000 ml?

3 A random sample of 12 hollyhock plants, grown from the seeds in a particular packet, was taken, and the height of each plant was measured, in metres. The results are summarized by $\sum x = 28.43$, $\sum (x - \bar{x})^2 = 28.47$, where $\bar{x}$ is the sample mean.

Making a suitable assumption about the distribution of heights, which should be stated, find a 90% confidence interval for the mean height of hollyhock plants grown from that packet.

4 Ten red apples fall off the display in a supermarket. The weights (in grams) of the fallen apples are summarized by $\sum x = 1023.7$, $\sum(x - \bar{x})^2 = 274.96$, where $\bar{x}$ is the sample mean.

Treating the fallen apples as being a random sample, determine a 99% confidence interval for the mean weight of a red apple, stating any assumptions that you have made.

5 The weights (in grams) of a random selection of pebbles from those on a particular beach were as follows:

103.2, 81.1, 109.4, 133.2, 181.1, 142.9

Regarding these values as being independent observations from a normal population, obtain a 95% confidence interval for the mean weight of the pebbles on the beach.

6 In a test match, the demon bowler Fasta Ami is bowling. The speed at which the cricket ball leaves his hand is measured (in mph) for a random selection of 10 deliveries. The results are summarised by $\sum x = 887.9$, $\sum(x - \bar{x})^2 = 332.2$, where x is the speed of a delivery and $\bar{x}$ is the sample mean.

a) Assuming that these results are observations from a normal distribution, obtain unbiased estimates of the mean and variance of this distribution, and hence obtain a 99% confidence interval for the mean.

b) Assuming that the distribution has mean and variance equal to the unbiased estimates obtained from this sample, determine the probability that Fasta delivers a ball at more than 90 mph.

7 The lengths (in cm) of the stems of a sample of a certain type of daffodil were as follows:

60.2	63.5	56.2	59.2	68.7	70.2
73.4	55.1	67.2	63.4	62.6	61.2

Use the sample data to obtain a 99% confidence interval for the mean stem length of this type of daffodil. State any assumptions you have made in obtaining your confidence interval.

Explain carefully the meaning of 99% *confidence* as applied to an interval in this context.

Summary

You should now be able to ...	Check out
1 Use tables of the t-distribution.	**1** a) If X has a t_6-distribution, find c, where $P(X < c) = 0.99$. b) If X has a t_{12}-distribution, find c, where $P(X > c) = 0.05$.
2 Determine a confidence interval for the mean of a normal population with unknown variance.	**2** The random variable X has a normal distribution. Sixteen observations in a random sample from this distribution are summarized by: $\sum x = 182.4$, $\sum(x - \bar{x})^2 = 1422.48$, where $\bar{x}$ denotes the sample mean. Determine a 95% confidence interval for the mean of the distribution of X.

Revision exercise 4

1 As part of a quality control procedure, the manager of a bottling plant regularly takes random samples of 20 bottles from the production line and measures the contents, x cl, of each bottle. On one occasion, he finds that:

$$\sum x = 1518.9 \text{ and } \sum(x - \bar{x})^2 = 7.2895$$

where $\bar{x}$ is the sample mean.

Assuming that the contents are normally distributed, find a 95% confidence interval for the mean. (*AQA, 2002*)

2 Charles is an athlete specialising in throwing the javelin. During practice, he throws the following distances, in metres.

40.3 39.8 41.6 42.8 39.0 38.6 40.8 41.1

a) Calculate unbiased estimates of the mean and the variance of the distance thrown.

b) i) Hence calculate a 95% confidence interval for the mean distance thrown.

ii) State **two** assumptions that you need to make in order to do this calculation.

(AQA, 2003)

3 The flight times, x minutes, from Airport A to Airport B were recorded on 14 occasions during a particular week and it was found that:

$$\sum x = 1337 \quad \text{and} \quad \sum(x - \bar{x})^2 = 366.5$$

where $\bar{x}$ is the sample mean. The flight times may be assumed to be a random sample from a normal distribution.

a) Calculate unbiased estimates of the mean, μ, and the variance, σ^2, of the flight times.

b) Calculate a 95% confidence interval for μ.

(AQA, 2003)

4 Emma is learning to cook and wants to find a quick way of measuring out flour. Her father tells her that a heaped tablespoon of flour weighs 50 grams.

To investigate this statement, Emma spoons out one heaped tablespoon of flour and tips the flour onto a piece of paper. She carries out this process a total of nine times, tipping each tablespoon of flour onto a separate piece of paper. She then weighs each measure of flour separately and records its weight, x grams.

Emma's results are summarised below.

Sample size = 9 $\sum x = 473$ $\sum(x - \bar{x})^2 = 76.2222$

where $\bar{x}$ is the sample mean.

a) Explain why Emma spooned out all the measures of flour and then weighed them, rather than weighing each measure as soon as she had spooned it out.

b) Calculate unbiased estimates for the mean, μ, and the variance, σ^2, of the weight of a heaped tablespoon of flour spooned out by Emma.

c) Stating the necessary distributional assumption, calculate a 90% confidence interval for the mean, μ.

d) Use the confidence interval you found in part c) to comment on the usefulness, for Emma, of this method of measuring out flour.

(AQA, 2004)

5 The contents of jars of honey may be assumed to be normally distributed. The contents, in grams, of a random sample of eight jars were as follows:

458 450 457 456 460 459 458 456

a) Calculate a 95% confidence interval for the mean contents of all jars.

b) On each jar it states 'Contents 454 grams.' Comment on this statement using the given sample and your results in part a).

c) Given that the mean contents of all jars is 454 grams, state the probability that a 95% confidence interval calculated from the contents of a random sample of jars will **not** contain 454 grams. (*AQA, 2001*)

6 A firm is considering providing an unlimited supply of free bottled water for employees to drink during working hours. To estimate how much bottled water is likely to be consumed, a pilot study is undertaken. On a particular day-shift, ten employees are provided with unlimited bottled water. The amount each one consumes is monitored. The amounts, in ml, consumed by these ten employees are as follows:

110 0 640 790 1120 0 0 2010 830 770

a) Assuming the data may be regarded as a random sample from a normal distribution, calculate a 95% confidence interval for the mean amount consumed on a day-shift.

b) i) Give a reason, based on the data collected, why the normal distribution may not provide a suitable model for the amount of free bottled water which would be consumed by employees of the firm.

ii) A normal distribution may provide an adequate model but cannot provide an exact model for the amount of bottled water consumed. Explain this statement giving a reason which does not depend on the data collected.

c) Following the pilot study the firm offers free bottled water to all the 135 employees who work on the night-shift. The amounts they consume on the first night have a mean of 960 ml with a standard deviation of 240 ml.

i) Assuming that these data may be regarded as a random sample, calculate a 90% confidence interval for the mean amount consumed on a night-shift.

ii) Explain why it was **not** necessary to know that the data came from a normal distribution in order to calculate the confidence interval in part c) i).

iii) Give **one** reason why it may be unrealistic to regard the data as a random sample of the amounts that would be consumed by all employees if the scheme was introduced on all shifts on a permanent basis. (*AQA, 2002*)

7 A health food co-operative imports a large quantity of dates and packs them into plastic bags labelled '500 grams'. George, a Consumer Protection Officer, checked a random sample of the bags and found the weights, in grams, of the contents were as follows:

497 501 486 502 492 508 489 494

a) Assuming that the weights follow a normal distribution, calculate, for the mean weight of contents of all the bags:
   i) a 95% confidence interval;
   ii) an 80% confidence interval.

b) George is uncertain what conclusion to draw and so decides to start again by taking a much larger sample of the bags. He takes a random sample of 95 bags and finds they have a mean weight of 499.6 grams and a standard deviation of 9.3 grams.

   Use this larger sample to calculate a 90% confidence interval for the mean weight of contents of all the bags.

c) The health food co-operative also imports raisins. George intends to take a random sample of 500-gram packets of raisins, weigh the contents and use the results to calculate an 80% and a 95% confidence interval for the mean weight, μ, of the contents of all the co-operative's packets of raisins.
   i) Find the probability that:
      A) the 80% confidence interval contains μ;
      B) the 95% confidence interval contains μ but the 80% confidence interval does not.
   ii) Instead of calculating both confidence intervals from the same sample, George now decides to calculate the 95% confidence interval from one sample and the 80% confidence interval from a second independent random sample. Find the probability that the 95% confidence interval contains μ but the 80% confidence interval does not.

(AQA, 2003)

8 Applicants to join a police force are tested for physical fitness. Based on their performance, a physical fitness score is calculated for each applicant. Assume that the distribution of scores is normal.

a) The scores for a random sample of ten applicants were:

   55 23 44 69 22 45 54 72 34 66

   Calculate a 99% confidence interval for the mean score of all applicants.

b) The scores of a further random sample of 110 applicants had a mean of 49.5 and a standard deviation of 16.5.

   Use the data from this second sample to calculate:
   i) a 95% confidence interval for the mean score of all applicants;
   ii) an interval within which the score of approximately 95% of applicants will lie.

c) By interpreting your results in parts b) i) and b) ii), comment on the ability of the applicants to achieve a score of 25.

d) Give **two** reasons why a confidence interval based on a sample of size 110 would be preferable to one based on a sample of size 10.

e) It is suggested that a much better estimate of the mean physical fitness of all recruits could be made by combining the two samples before calculating a confidence interval. Comment on this suggestion. *(AQA, 2003)*

9 A company manufactures components for the motor industry. The components are designed to have a length of 135.0 mm. Each day a technician takes a random sample of components and calculates a 95% confidence interval for the mean length.

a) On a particular Monday the technician takes a sample of nine components. Their lengths, in millimetres, are as follows:

135.1 135.7 134.9 135.2 136.3 135.7 135.9
136.0 135.6

Assuming the lengths of components produced on this Monday may be modelled by a normal distribution, calculate a 95% confidence interval for the mean length. Give your answer to an appropriate degree of accuracy.

b) On a particular Friday a random sample of 60 components had a mean length of 135.8 mm and a standard deviation of 3.9 mm. Calculate a 95% confidence interval for the mean length of components produced on this Friday.

c) The company overhauls all machines when the confidence interval calculated does not contain 135.0 mm. State whether increasing the sample size, n, is likely to increase, decrease or leave unaltered:
i) the width of the confidence interval,
ii) the frequency of machine overhauls, explaining your answer.

d) Using your results in parts a) and b) as an example, or otherwise, explain why it is possible that production on a day when the confidence interval does not contain 135.0 mm will sometimes be more satisfactory than production on a day when the confidence interval does contain 135.0 mm. *(AQA, 2004)*

5 Hypothesis testing

This chapter will show you how to

- ✦ Understand the concepts of null and alternative hypotheses
- ✦ Distinguish between one-tailed and two-tailed tests
- ✦ Understand the concepts of Type I errors and Type II errors
- ✦ Understand and apply the concepts of significance level, critical value, critical region and test statistic
- ✦ Test for the mean of a normal population with known variance
- ✦ Test for the mean of a normal population with unknown variance
- ✦ Test for the mean of a population using a large sample and a normal approximation

Before you start

You should know how to ...	Check in
1 Standardize a normal random variable.	**1** The random variable X has a normal distribution with mean 8 and variance 4. Express the statement $\mathrm{P}(X > 9)$ as a statement about the standard normal random variable, Z.
2 Use Table 4 (in the Appendices) to find values of z for a given value of p or $1 - p$.	**2** Use Table 4 (in the Appendices) to find the value of z such that: a) $\mathrm{P}(Z < z) = 0.95$ b) $\mathrm{P}(Z > z) = 0.025$ c) $\mathrm{P}(Z < z) = 0.005$ d) $\mathrm{P}(\lvert Z \rvert > z) = 0.10$
3 Calculate an unbiased estimate of a population mean.	**3** Calculate an unbiased estimate of the mean of X in each of the following cases: a) A random sample of nine observations of X is: 1.1, 4.2, 8.3, 6.4, 7.2, 5.3, 4.9, 5.1, 7.0 b) A random sample of 65 observations on X has $\sum x = 32\,851$. c) A random sample of 100 observations on X is summarized by: x: 10– 20– 30– 50– 75–100 Frequency: 9 25 37 18 11
4 Calculate an unbiased estimate of a population variance.	**4** Calculate an unbiased estimate of a population variance: a) for the data in question **3** a). b) for the data in question **3** b), with $\sum(x - \bar{x})^2 = 5184$, where $\bar{x}$ is the sample mean. c) for the data in question **3** c).

One wet Saturday morning, having nothing better to do, you decide to weigh some tins of the extraordinarily cheap *Ripoff* baked beans that you bought the previous week. To your amazement, the twelve tins have an average weight that is 10 g less than the weight stated on the tins. Should you report the manufacturers to the authorities? In despair you eat the baked beans (and learn about hypothesis tests).

5.1 Null and alternative hypotheses

Hypothesis tests are also called **significance tests**.

In a hypothesis test, two hypotheses (called the 'null hypothesis' and the 'alternative hypothesis') are compared. The outcome of the test will be that one of these hypotheses is 'accepted' and the other will be 'rejected'.

Usually the **null hypothesis** specifies *a particular value* for some population parameter whereas the **alternative hypothesis** specifies *a range of values* that excludes the value specified by the null hypothesis. Here are some examples:

S2

In this chapter, only hypotheses concerning the mean are considered.

Parameter	*Null hypothesis*	*Alternative hypothesis*
Mean, μ	$\mu = 435$	$\mu \neq 435$
Mean, μ	$\mu = 435$	$\mu < 435$

To save writing out 'null hypothesis' and 'alternative hypothesis' lots of times, the hypotheses are denoted by H_0 and H_1, respectively. Thus the first pair of hypotheses in the table would become:

$$H_0: \mu = 435 \qquad H_1: \mu \neq 435$$

All the hypotheses tested in this chapter concern the population mean μ. In each case the test is based on the distribution of $\overline{X}$, the sample mean, with the test statistic being a function of the observed $\bar{x}$. When the population variance σ^2 is unknown, the unbiased estimate, s^2, is used in its place. In this chapter the following situations are considered:

Population	Variance	Sample size	Test statistic	Reference distribution
Normal	Known	Any	$\dfrac{\bar{x}-\mu}{\frac{\sigma}{\sqrt{n}}}$	N(0, 1)
Normal	Unknown	Any	$\dfrac{\bar{x}-\mu}{\frac{s}{\sqrt{n}}}$	t_{n-1}
Any	Known	$\geqslant 30$	$\dfrac{\bar{x}-\mu}{\frac{\sigma}{\sqrt{n}}}$	N(0, 1)
Any	Unknown	$\geqslant 30$	$\dfrac{\bar{x}-\mu}{\frac{s}{\sqrt{n}}}$	N(0, 1)

In each case the null hypothesis specifies a value, μ_0, for the unknown mean. Using this value, the probabilities of events of interest (such as the sample mean being greater than 450) can be determined. Knowing these probabilities enables the development of rules for deciding which hypothesis to accept.

Example 1

Ten independent observations are to be taken from a $N(\mu, 40)$ distribution. The hypotheses are $H_0 : \mu = 20$, $H_1 : \mu > 20$. The following procedure has been proposed:

'Reject H_0 (and accept H_1) if $\overline{X} > 23.29$
Accept H_0 otherwise'.

Assuming H_0 is true, determine the probabilities of accepting and rejecting H_0 when using this procedure.

Assuming that $\mu = 20$ the distribution of $\overline{X}$ is $N\left(20, \frac{40}{10}\right) = N(20, 4)$, so that:

$$\frac{\overline{X} - 20}{2} \sim N(0, 1)$$

Thus:

$$P(\overline{X} > 23.29) = P\left(Z > \frac{23.29 - 20}{2}\right) = P(Z > 1.645)$$

where $Z \sim N(0, 1)$. This tail probability is 5%. Hence, if $\mu = 20$, the probabilities of accepting and rejecting H_0, when using the procedure, are 95% and 5%, respectively.

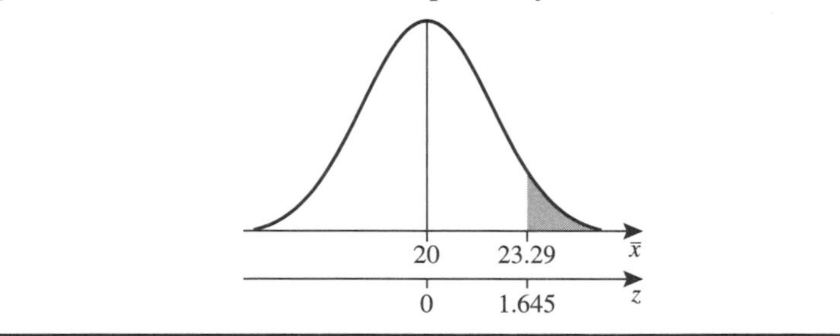

In English law the prisoner in the dock is treated as being innocent until, after hearing the evidence, the prisoner has been judged guilty. In the same way, after examining the evidence (the data), the null hypothesis is accepted unless the evidence suggests that, compared to the the alternative hypothesis, it is implausible. Thus acceptance of the null hypothesis does not prove that it is correct, any more than the verdict 'Not guilty' proves that the prisoner is innocent.

Identifying the two hypotheses

It is often easy to recognise a question on hypothesis tests, because the word 'test' appears in the question! However, it can sometimes be difficult to identify the two hypotheses.

<u>The null hypothesis</u>
This states that a parameter has some precise value:

1. The value that occurred in the past.
2. The value claimed by some person.
3. The (target) value that is supposed to occur.

It is often necessary to assume that σ^2 is unchanged from some previous value.

The alternative hypothesis

The alternative hypothesis involves the use of one of the signs $>$, $<$ or $\neq$. A decision has to be made as to which is appropriate. Generally, exam questions attempt to signal which sign is to be used by means of suitable phrases:

$\neq$ 'change', 'different', 'affected';
$>$ or $<$ 'less than', 'reduced', 'better', 'increased', 'overweight'.

Most questions use words like these, and you should always choose the alternative hypothesis that is appropriate to the problem.

Exercise 5A

In each of questions **1** to **5** write down appropriate null and alternative hypotheses.

1 A machine is supposed to place 1 kg of flour in each bag. It is suspected that the controls may be incorrectly set. The weights of a random sample of 25 bags are recorded.

2 On average, in a notorious inner city ring road, the average speed at 10 pm is 41.0 mph. The Chief Constable of the district launches a high-profile police initiative designed to reduce speeding. During the next week the speeds of a random sample of 100 cars are recorded.

3 The average weight of a colourful sweet is 1 g. The manufacturer brings out a limited edition of the sweet in a new colour. Billy Bunter buys a random sample of the new sweets and weighs each sweet (before eating it).

4 For many years a teacher has been struggling to teach Statistics, using the same textbook. Her students obtained an average mark of 51.2 in school exams. This year a new Statistics textbook appears which claims to be easier to understand and thus to result in the students doing better in their exams.

5 Farmer Giles grows potatoes. His average annual crop (per plant) is 2.4 kg. One year Harry Hayseed suggests that Farmer Giles should use a new fertiliser. The farmer uses this fertiliser and records the crops that he obtains.

S2

5.2 The language of hypothesis tests

In addition to the terms 'Null hypothesis' and 'Alternative hypothesis', there are a number of other terms and concepts that are required.

- **Test statistic**
  A test statistic is a function of the data (or the corresponding random variables) that is used to decide between the two hypotheses. For hypotheses concerning the population mean, the test statistic involves $\bar{x}$ (or $\bar{X}$), the sample mean.
- **The critical region**
  The set of values of the test statistic that leads to the rejection of H_0 in favour of H_1 is called the **critical region**.
- **Significance level**
  When the population parameter has the value specified by H_0, the probability that H_0 is nevertheless rejected in favour of H_1 is

Another name for the critical region is the **rejection region**. The set of values that leads to the acceptance of H_0 is then referred to as the **acceptance region**.

called the **significance level**. Thus the significance level is the probability, when the null hypothesis is correct, of obtaining a value of the test statistic in the critical region and thereby rejecting the null hypothesis.

In Example 1 (Page 114) the significance level was 5% and the critical region was the set $\bar{x} > 23.29$, or, equivalently, $z > 1.6449$.

If the value of the test statistic falls in the critical (rejection) region then the outcome is said to be '**significant**'. If the significance level were $\alpha\%$, then the outcome would be described as being '**significant at the $\alpha\%$ level**'.

In Example 1 (Page 114), 23.29 was a critical value for $\bar{x}$.

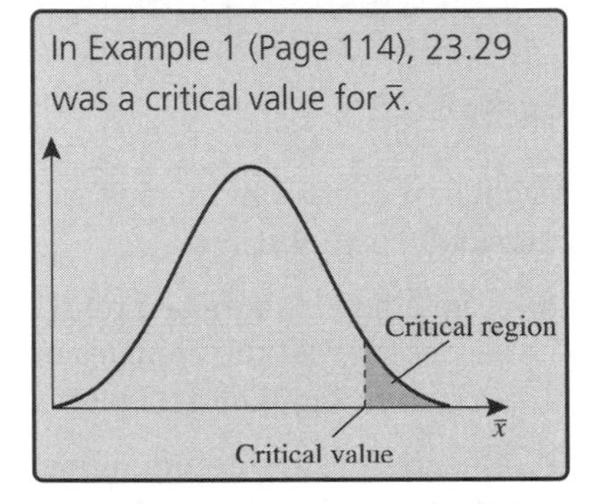

- **Critical value**
  A value that determines a limit of the critical region is called a **critical value**.
- **One tailed and two tailed tests**
  Hypothesis tests in which H_1 involves either a '>' sign (as in Example 1) or a '<' sign are called **one tailed tests**. The critical regions in these cases involve values in the corresponding tail of the distribution specified by H_0.
  Hypothesis tests in which H_1 involves a '$\neq$' sign are called **two tailed tests**. In these cases the 'critical region' actually consists of two regions – one in each tail of the distribution specified by H_0. For an $\alpha\%$ significance level, there will be a probability of $\frac{1}{2}\alpha\%$ assigned to each tail.
  Three examples of critical regions (with probabilities shown shaded) are illustrated below for the case of a single observation on a random variable X having a normal distribution with variance 1.
  As usual, it is convenient to work with Z, where $Z = \dfrac{X - \mu}{\sigma}$, and so both x and z scales are shown. The z-values, given in the diagrams to two decimal places, are taken from Table 4 of the AQA tables, which are reproduced in the Appendices to this book.
  The first example illustrates a one tailed test, while the other two examples show two tailed tests.

Smaller significance levels result in smaller critical regions. By custom, the values most used are 5% and 1%.
An outcome that is significant at the $\alpha\%$ level is also significant at the $\beta\%$ level, for all $\beta > \alpha$. For example, an outcome that is significant at the 1% level is always significant at the 5% level.

One tailed and two tailed tests are sometimes referred to as **one-sided** and **two-sided** tests.

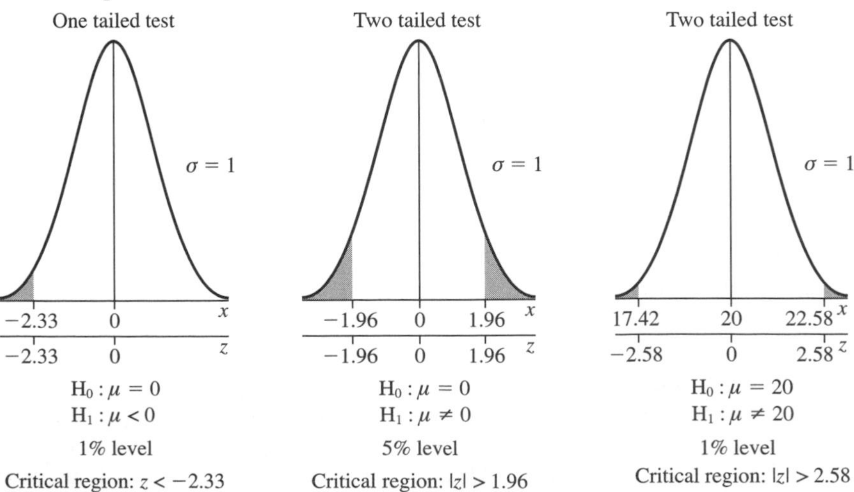

Although both one tailed tests and two tailed tests may appear in examination questions, some statisticians question whether one tailed tests should ever be used in practice. To see why this is so, consider the following situation:

> 'The mean breaking strength of a type of climbing rope is 200 kg. Scientists make an adjustment to the method of construction which, they claim, will result in an increase in the breaking strength. A random sample of 12 pieces of the new rope are tested ... .'

This appears very straightforward, with the hypotheses being:

H_0: $\mu = 200$ kg
H_1: $\mu > 200$ kg

However, suppose that the 12 pieces of new rope have the following breaking strengths:

187, 196, 193, 187, 194, 193, 197, 194, 191, 195, 194, 199

The hypothesis H_0 will certainly not be rejected in favour of H_1, since the mean breaking strength of the new rope appears to be about 193 or 194, and it is certainly not greater than 200. Since H_0 is not rejected, this means that it must be accepted (even though it is probably false; this is because the possibility that $\mu < 200$ kg has not been considered).

Acceptance of a null hypothcsis docs not mean that it is true; it only means that the data provide no reason to prefer the alternative hypothesis.

✦ Type I and Type II errors

The statistician's life is not a happy one! When conducting hypothesis tests there are two types of error that may occur, as the table below shows.

		Decision made	
		Accept H_0	Accept H_1
Reality	H_0 correct	Correct!	TYPE I ERROR
	H_1 correct	TYPE II ERROR	Correct!

When the mean has the value specified by H_0, a Type I error occurs whenever the observed value of the test statistic falls in the critical region.

As the table shows:

A **Type I error** occurs if, as a result of a test, H_0 is rejected in favour of H_1, even though H_0 is correct.

To reduce the probability of Type I errors, significance levels greater than 5% are rarely used in practice.

The probability of this error is controlled since:

$$P(\text{Type I error}) = \text{Significance level} \qquad (5.1)$$

By contrast:

A **Type II error** occurs if, as a result of a test, H_0 is accepted, even though it is incorrect.

The probability of a **Type II error** is not fixed, since it depends upon the extent to which the value of μ deviates from the value μ_0 specified by H_0: the probability is large if μ is close to μ_0, but small if μ is far from μ_0.

Example 2

S2

Bags of salt are supposed to weigh 30 kg. A change is made to the manufacturing process. The change is not supposed to affect the weights of the bags of salt. After the change to the process, a random sample of 25 bags is selected and weighed.

a) State suitable null and alternative hypotheses.

b) Explain, in context, what would constitute:
i) a Type I error ii) a Type II error.

a) Denoting the population mean weight after the change by μ kg, a suitable pair of hypotheses are:

H_0: $\mu = 30$
H_1: $\mu \neq 30$

b) i) In context, a Type I error would occur if the hypothesis that the bags come from a population with mean 30 kg was rejected when, in fact, it was true.

ii) A Type II error would occur if it was accepted that the mean weight was 30 kg, when, in fact, this was not the case.

Exercise 5B

In each of questions 1 to 5 write down what, in the context of the question, would constitute: a) a Type I error b) a Type II error.

1 A machine is supposed to place 1 kg of flour in each bag. It is suspected that the controls may be incorrectly set. The weights of a random sample of 25 bags are recorded.

2 On average, in a notorious inner city ring road, the average speed at 10 p.m. is 41.0 mph. The Chief Constable of the district launches a high-profile police initiative designed to reduce speeding. During the next week the speeds of a random sample of 100 cars are recorded.

3 The average weight of a colourful sweet is 1 g. The manufacturer brings out a limited edition of the sweet in a new colour. Billy Bunter buys a random sample of the new sweets and weighs each sweet (before eating it).

4 For many years a teacher has been struggling to teach Statistics, using the same textbook. Her students obtained an average mark of 51 in school exams. This year a new Statistics textbook appears which claims to be easier to understand and thus to result in the students doing better in their exams.

5 Farmer Giles grows potatoes. His average annual crop (per plant) is 2.4 kg. One year Harry Hayseed suggests that Farmer Giles should use a new fertiliser. The farmer uses this fertiliser and records the crops that he obtains.

The general test procedure

There are six steps:

1. Write down the two hypotheses, for example, H_0: $\mu = \mu_0$ and H_1: $\mu > \mu_0$, where μ_0 is a specified value of μ.
2. Identify an appropriate test statistic and the distribution of the corresponding random variable (using the parameter value specified by H_0).
3. Identify the significance level. This is P(Type I error).
4. Determine the critical region.

   Now collect the data

5. Calculate the value of the test statistic.
6. Determine the outcome of the test.

Remember that the hypotheses concern the value of the population mean, *not* the sample mean.

In an examination, in stating the outcome of the test, it is not sufficient to state 'H_0 is accepted (or rejected)' – you should also state what this means in the context of the question.

It is important to decide upon the hypotheses and the significance level (and hence the critical region) *before* looking at the actual data so as not to be accidentally biased. This avoids claims that the critical region has been carefully selected so as to get a desired result.

All too often lay observers (not, of course, well-trained statisticians!) choose their critical regions *after* seeing the data. Their 'significant' results must then be treated with caution.

Note that sometimes the null hypothesis may not appear to refer to a precise value:

> 'It is claimed that a new rope brought onto the market has a breaking strength in excess of 200 kg. A random sample of 12 pieces of the new rope are tested ... .'

Here it appears that the hypotheses are:

$$\mu \leqslant 200 \text{ kg} \quad \text{and} \quad \mu > 200 \text{ kg}$$

In order to see how to proceed, consider two alternative specific null hypotheses such as:

$$H_0': \mu = 200 \text{ kg}$$

and

$$H_0'': \mu = 190 \text{ kg}$$

Consider H_0' and suppose that the outcome of the test is that H_0' is rejected. This only occurs if $(\bar{x} - 200)$ is unacceptably large. However, since $(\bar{x} - 190)$ will then be larger than $(\bar{x} - 200)$, this too must be unacceptably large. Thus H_0'' would also have been rejected. The same argument would apply for any value of μ less than 200 kg. All the possible cases are therefore covered by using:

H_0: $\mu = 200$ kg
H_1: $\mu > 200$ kg

5.3 Test for the mean of a normal distribution with known variance

Suppose that the variance of the normal population is known to be σ^2, and that the null hypothesis specifies that $\mu = \mu_0$. Assuming the null hypothesis, the distribution of $\overline{X}$, the random variable corresponding to the mean of a random sample of n observations from the population, is:

S2

$$N\left(\mu_0, \frac{\sigma^2}{n}\right)$$

Example 3

The length of string in Stringo balls of string has a normal distribution with mean μ metres and standard deviation 0.15 metres. On the packaging it is claimed that $\mu = 3$. A trading officer examines a random sample of sixteen balls. He finds that these have a mean length of 2.94 metres. Test whether this provides evidence, at the 5% significance level, that the mean length has been overstated.

1. *Write down H_0 and H_1.*
   The question is whether the mean is as great as the value claimed, so the test is one tailed:

   H_0: $\mu = 3$
   H_1: $\mu < 3$

2. *Identify an appropriate test statistic and the distribution of the corresponding random variable (using the parameter value specified by H_0).*
   Since the distribution being sampled is a normal distribution, the distribution of $\overline{X}$ is normal. Since $\sigma = 0.15$ and $n = 16$, an appropriate test statistic is given by:

   $$z = \frac{\bar{x} - 3.0}{\frac{0.15}{\sqrt{16}}} = \frac{(\bar{x} - 3)}{0.0375}$$

   If $\mu = 3$ (as specified by H_0), z is an observation from a standard normal distribution.

3. *Identify the significance level.*
   The question specifies 5%.

When a test is based on the N(0, 1) distribution, it is often referred to as a 'z-test', with the test statistic being described as the z-statistic.

For a z-test based on a sample from a normal population with known variance, the sample size, n, does not need to be large.

4. *Determine the critical region.*
The test is one tailed.
Since $P(Z < -1.6449) = 0.05$, an appropriate procedure is to reject H_0 in favour of H_1 if and only if z is less than -1.6449.

5. *Calculate the value of the test statistic.*
Since $\bar{x} = 2.94$,
$$z = \frac{(2.94 - 3)}{0.0375} = -1.6$$

6. *Determine the outcome of the test.*
Since z is (just) greater than the critical value (-1.6449), H_0 is accepted.
The mean length of string in a ball is accepted as being 3 metres.

If the value of z falls just outside the critical region then the null hypothesis is accepted. However, it would be as well to note that the outcome was nearly significant. A series of 'nearly significant' results concerning the same parameter would probably be regarded as significant.

S2

Calculator practice

Some calculators can perform a z-test with just a few key strokes. Typically the sequence will involve switching to a special mode and then selecting from the options available. Some machines can accept either the individual data values or the sample mean. In addition you will be required to provide the values of σ^2 and μ_0. The calculator then provides the value of z.

If your calculator has these capabilities, then it may also prompt for information about the alternative hypothesis, in which case, as well as the value of z, it may calculate a '**p-value**'. The p-value is the probability of observing a result as extreme, or more extreme, given the null hypothesis. The probability quoted depends upon the alternative hypothesis:

Alternative hypothesis	Probability calculated to give p-value
$H_1: \mu < \mu_0$	$P(Z < z)$
$H_1: \mu > \mu_0$	$P(Z > z)$
$H_1: \mu \neq \mu_0$	$P(\lvert Z \rvert > \lvert z \rvert)$

If the calculated p-value is less than 0.05, then there is significant evidence to reject H_0 in favour of H_1 at the 5% level. If the calculated p-value is less than 0.01, then there is significant evidence to reject H_0 in favour of H_1 at the 1% level, and so on.

In Example 3 the alternative hypothesis was $H_1: \mu < 3$, the p-value given by a calculator will be $P(Z < z)$, with, in this case, $z = -1.6$.

Since $P(Z < -1.6) = 0.0548$, and $f = 0.0548 > 0.05$, there is no significant evidence to reject H_0 in favour of H_1, and H_0 is therefore accepted.

When using a calculator, always check that you have entered the data correctly!

Remember that just quoting a z-value (or a p-value) will not be sufficient to earn full marks on a question.

Using p-values it is not necessary to determine the critical region (unless required to do so by the question).

Example 4

A random sample of 36 observations is to be taken from a normal distribution with variance 100. In the past the distribution has had a mean of 83.0, but it is believed that the mean may recently have changed.

a) Using a 5% significance level, determine an appropriate test of the null hypothesis, H_0, that the mean is 83.0.

b) When the sample is actually taken it is found to have a mean of 86.2. Does this provide significant evidence against H_0?

c) Suppose it is known that, if the population mean has changed, then it can only have increased. How would this knowledge affect the conclusions?

S2

a) 1. *Write down H_0 and H_1.*
There is no suggestion in the initial question that any change can only be in one direction. The test is therefore two tailed:

$$H_0: \mu = 83$$
$$H_1: \mu \neq 83$$

2. *Identify an appropriate test statistic and the distribution of the corresponding random variable (using the parameter value specified by H_0).*
Since the distribution being sampled is a normal distribution, the distribution of $\overline{X}$ is normal. Since $\sigma = 10$ and $n = 36$, an appropriate test statistic is:

$$z = \frac{\bar{x} - 83.0}{\frac{10}{\sqrt{36}}} = 0.6(\bar{x} - 83.0)$$

If $\mu = 83$ (as specified by H_0), z is an observation from a standard normal distribution.

3. *Identify the significance level.*
The question specifies 5%.

4. *Determine the critical region.*
The test is two tailed.
Since $P(Z > 1.9600) = 0.025$, and $P(Z < -1.9600) = 0.025$, an appropriate procedure is to accept H_0 if z lies in the interval $(-1.9600, 1.9600)$ and otherwise to reject H_0 in favour of H_1.
Thus the critical values are ± 1.9600 and the critical region is $|z| > 1.9600$.

With two tailed tests, the critical region is usually taken to be symmetrical about μ_0. For a 5% significance level this implies 2.5% in each tail.

The critical value 1.9600 is the value given in Table 4 (in the Appendices) corresponding to a cumulative probability of 0.975 for a standard normal random variable.

b) 5. *Calculate the value of the test statistic.*
Since $\bar{x} = 86.2$, $z = 1.92$.

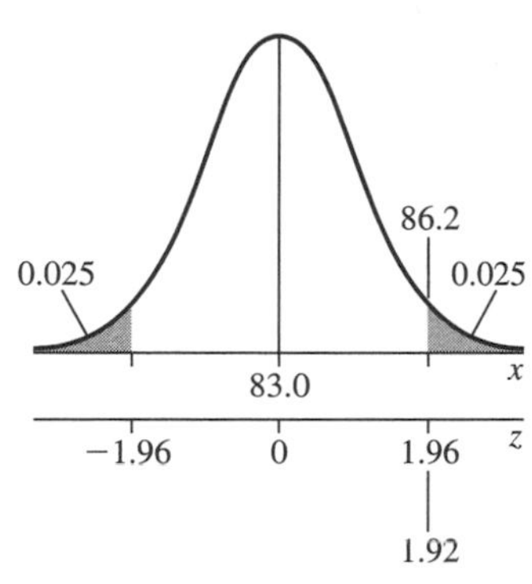

6. *Determine the outcome of the test.*
Since $-1.96 < z < 1.96$, H_0 is accepted.

In other words, there is no significant evidence, at the 5% level, that the mean has changed from its previous value of 83.0.

In this case the p-value ($= P(|Z| > 1.92)$) is equal to 0.0549. Since $0.0549 > 0.05$, there is no significant evidence, at the 5% level, to reject H_0, which is therefore accepted.

This does *not* imply that the mean is unchanged, it is simply that the mean of this particular sample did not happen to give a value of z falling in the critical region.

S2

c) If it is known that the population mean cannot have decreased then H_0 will be rejected in favour of H_1 only if $\bar{x}$ is unusually large. The test is now one-tailed with H_1: $\mu > 83.0$. Since $P(Z > 1.6449) = 0.05$, an appropriate procedure is now to reject H_0 in favour of H_1 if z is greater than 1.6449.

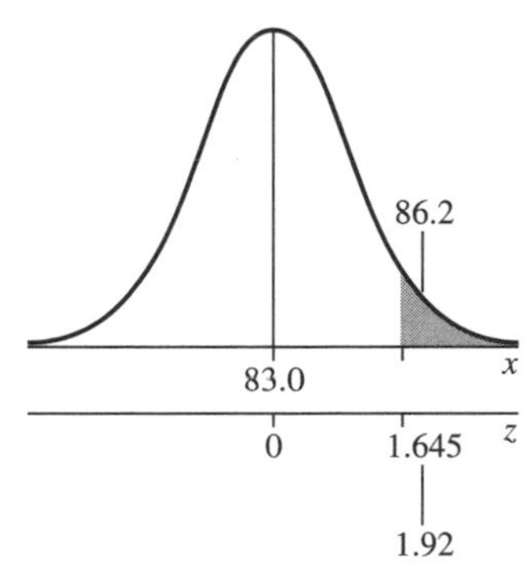

In this case the p-value ($= P(Z > 1.92)$) is equal to 0.0274. Since $0.0274 < 0.05$, H_0 is rejected in favour of H_1.

Since 1.92 *is* greater than 1.6449, H_0 is rejected and the alternative hypothesis is accepted in its place.

There is significant evidence, at the 5% level, that the population mean has increased from its previous value.

This result is not the contradiction that it appears to be! The test in part b) failed to reject H_0, but, as previously stated, if a test fails to reject H_0, that does *not* mean that H_0 is correct – it simply implies that, if H_0 is false, then the test has failed to detect the fact.

Exercise 5C

1 Jars of honey are filled by a machine. It has been found that the quantity of honey in a jar has a normal distribution with mean 460.3 g and standard deviation 3.2 g. It is believed that the controls have been altered in such a way that, although the standard deviation is unaltered, the mean quantity may have changed. A random sample of 60 jars is taken and the mean quantity of honey per jar is found to be 461.2 g. State suitable null and alternative hypotheses, and carry out a test using a 5% level of significance.

2 Observations of the time taken to test an electrical circuit board show that it has mean 5.82 minutes with standard deviation 0.63 minutes. As a result of the introduction of an incentive scheme, it is believed that the inspectors may be carrying out the test more quickly. It is found that, for a random sample of 10 tests, the mean time taken is 5.28 minutes. Assuming a normal distribution with an unchanged variance, carry out a test at the 5% significance level.

3 A light bulb manufacturer has established that the life of a bulb has a normal distribution with mean 95.2 days and standard deviation 10.4 days. Following a change in the manufacturing process which is intended to increase the life of a bulb without affecting the population standard deviation, a random sample of 96 bulbs will be taken. From this sample the mean life of a bulb will be determined, and a significance test will be carried out, at the 1% level.

a) Find the critical region for $\bar{x}$.

b) State the probability of a Type I error.

c) Explain, in the context of the question, what is meant by a 'Type II error'.

d) Determine the outcome of the test, given that the sample mean is 96.6 days.

4 The length of wire in the rolls of wire made by a particular manufacturer has mean μ m and variance 27.4 m^2. The manufacturer claims that $\mu = 200$. A random sample of 8 rolls of wire is taken and the sample mean is found to be 197.2 m. Assuming a normal distribution, test whether this provides significant evidence, at the 3% level, that the manufacturer's claim overstates the value of μ.

5 The distance driven by a van driver in a week is a normally distributed variable having mean 980 km and standard deviation 106 km. New driving regulations are introduced and, in the first 20 weeks after their introduction, he drives a total of 18 900 km. Assuming a normal distribution with an unchanged standard deviation, test, at the 10% level of significance, whether his mean weekly driving distance has changed.

6 In a large population of turkeys, the distribution of the weight of a turkey has a normal distribution with mean μ kg and standard deviation σ kg. A random sample of 100 turkeys is taken from the population. The mean weight (in kg) for the sample is denoted by $\overline{X}$.

a) State the approximate distribution of $\overline{X}$, giving its mean and standard deviation.

b) The sample values are summarised by $\sum x = 1086.91$, where x kg is the weight of a turkey. Given that, in fact, $\sigma = 1.71$, test, at the 1% level of significance, the null hypothesis $\mu = 10.5$ against the alternative hypothesis $\mu > 10.5$, stating whether you are using a one tailed or a two tailed test.

c) In the context of the question explain what is meant by:
i) a Type I error ii) a Type II error.

S2

5.4 Test for the mean of a normal distribution with unknown variance

When the variance of the normal population is unknown, the t-distribution must be used in place of the normal distribution.

The t-distribution was introduced in Chapter 4.

The test statistic is now t, defined by:

$$t = \frac{\bar{x} - \mu_0}{\sqrt{\frac{s^2}{n}}}$$

This test is generally referred to as the '***t*-test**', and the statistic as the '***t*-statistic**'.

where μ_0 is the population mean specified by the null hypothesis and s^2 (the unbiased estimate of the unknown population variance) is given by:

$$s^2 = \frac{1}{n-1}\sum(x - \bar{x})^2 = \frac{1}{n-1}\left\{\sum x^2 - \frac{(\sum x)^2}{n}\right\}$$

The random variable, T, corresponding to t, has a t_{n-1}-distribution.

Calculator practice

Some calculators can perform a t-test with just a few key strokes. Typically the sequence will involve switching to a special mode and then selecting from the options available. Some machines can accept either the individual data values or the sample size and sample mean and standard deviation. In addition you will be required to provide the value for μ_0. The calculator then provides the value of t.

If your calculator has these capabilities, then, as in the case of the z-test, it may prompt for information about the alternative hypothesis and calculate a 'p-value'. As for a z-test, the probability quoted depends upon the alternative hypothesis, but on this occasion the reference distribution will be the t_{n-1}-distribution.

When using a calculator always check that you have entered the data correctly.

Example 5

Bottles of wine are supposed to contain 75 cl of wine. An inspector takes a random sample of six bottles of wine and determines the volumes of their contents. Her results (in millilitres) were:

747.0, 751.5, 752.0, 747.5, 748.0, 748.0.

Assuming that the contents have a normal distribution, determine whether these results provide significant evidence, at the 5% level, that the population mean is less than 75 cl.

It is simplest to work in millilitres. The target quantity, 75 cl, is equivalent to 750 millilitres.

1. *Write down H_0 and H_1.*
   The test is one tailed. The hypotheses are:

   $$\mathrm{H}_0: \mu = 750$$
   $$\mathrm{H}_1: \mu < 750$$

2. *Identify an appropriate test statistic and the distribution of the corresponding random variable (using the parameter value specified by H_0).*
   Since the population is assumed to be normal, but with unknown variance, a t-test is appropriate. The t-statistic is:

   $$t = \frac{\bar{x} - 750}{\sqrt{\frac{s^2}{6}}}$$

   which, assuming H_0, is an observation from a t_5-distribution.

3. *Identify the significance level.*
   The question specifies 5%.

4. *Determine the critical region.*
   The test is one tailed. From Table 5 the upper 5% point of a t_5-distribution is 2.015, and hence, by symmetry, the lower 5% point is -2.015. This is therefore the critical value and the critical (rejection) region is $t < -2.015$.

   An appropriate procedure is therefore to accept H_0 if t is greater than -2.015 and otherwise to reject H_0 in favour of H_1.

5. *Calculate the value of the test statistic.*
   The data are summarised by $\sum x = 4494.0$ and $\sum x^2 = 3\,366\,029.50$, so that:

   $$s^2 = \frac{1}{5}\left\{3\,366\,029.50 - \frac{4494.0^2}{6}\right\} = 4.70$$

   Since $\bar{x} = \frac{4494.0}{6} = 749.0$,

   $$t = \frac{749.0 - 750}{\sqrt{\frac{4.70}{6}}} = -1.13$$

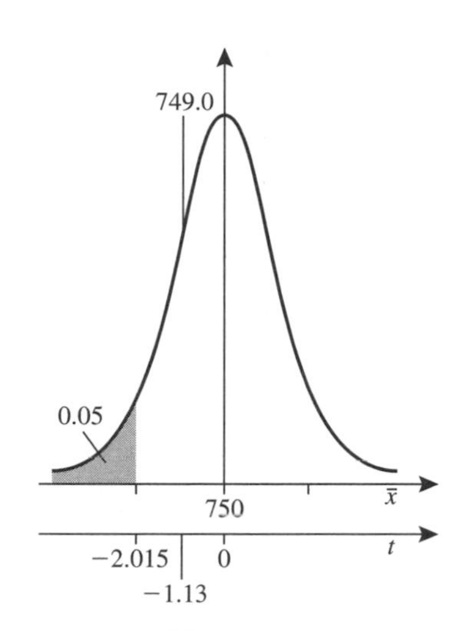

In this case the p-value is 0.155. Since $0.155 > 0.05$, the null hypothesis is accepted at the 5% significance level.

6. *Determine the outcome of the test.*
   Since $t > -2.015$, H_0 is accepted.

   There is no significant evidence, at the 5% level, that the bottles are being underfilled.

S2

Example 6

The amounts of curry powder in refill sachets have a normal distribution. The mean amount is claimed by the manufacturer to be 17 grams. The amounts of curry powder (in grams) in a random sample of ten sachets are:

16.9, 17.3, 17.1, 17.0, 16.5, 16.7, 16.6, 17.0, 16.4, 16.9

Test the manufacturer's claim at the 1% significance level.

1. *Write down H_0 and H_1.*
   The test is two tailed. The hypotheses are:

   H_0: $\mu = 17$
   H_1: $\mu \neq 17$

2. *Identify an appropriate test statistic and the distribution of the corresponding random variable (using the parameter value specified by H_0).*
   Since the population has a normal distribution, but with unknown variance, a t-test is appropriate. The t-statistic is:

$$t = \frac{\bar{x} - 17}{\sqrt{\dfrac{s^2}{10}}}$$

   which, assuming H_0, is an observation from a t_9-distribution.

For a t-test based on a sample from a normal distribution with unknown variance, the sample size n does not need to be large.

3. *Identify the significance level.*
   The question specifies 1%.

4. *Determine the critical region.*
   Since $n = 10$, a t_9-distribution is appropriate. For a two tailed 1% test, the relevant critical value is that in the column headed $p = 0.995$ and the row with $\nu = 9$ (this is 3.250). The critical region leading to the rejection of the null hypothesis that $\mu = 17$ is therefore $|t| > 3.250$.

5. *Calculate the value of the test statistic.*
   The data are summarised by $\sum x = 168.4$ and $\sum x^2 = 2836.58$, so that:

$$s^2 = \frac{1}{9}\left\{2836.58 - \frac{(168.4)^2}{10}\right\} = 0.0804$$

   Since $\bar{x} = \dfrac{168.4}{10} = 16.84$:

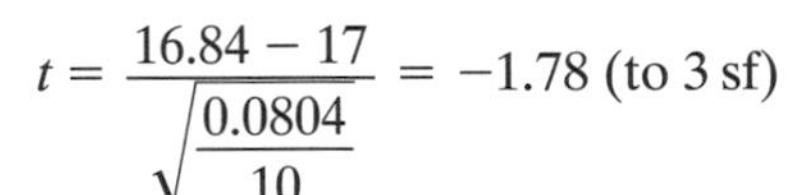

$$t = \frac{16.84 - 17}{\sqrt{\dfrac{0.0804}{10}}} = -1.78 \text{ (to 3 sf)}$$

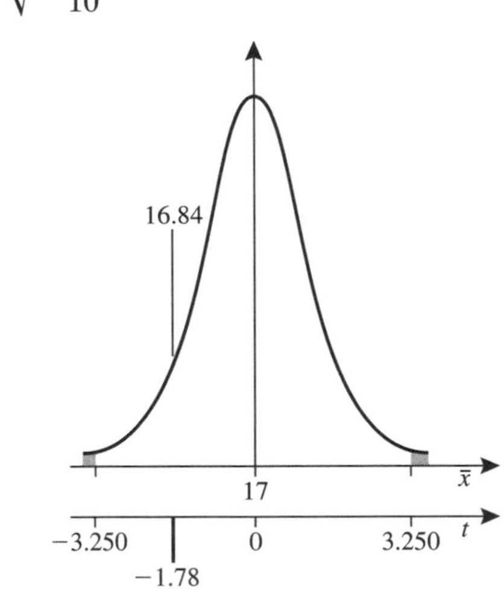

In this case the p-value is 0.108. Since $0.108 > 0.01$, H_0 is accepted.

6. *Determine the outcome of the test.*
   Since $-3.25 < -1.78 < 3.25$, H_0 is accepted.

   There is no significant evidence, at the 1% level, that the mean amount in a sachet is not 17 grams.

S2

Exercise 5D

1 The concentration of a substance is alleged to be 7 grams per litre. The concentrations in ten random samples are shown below.

Sample	1	2	3	4	5	6	7	8	9	10
Result (x)	2.1	9.0	15.1	5.2	8.6	5.6	10.7	10.4	2.2	6.1

These results are summarised by $\sum x = 75.0$, $\sum(x - \bar{x})^2 = 147.98$, where $\bar{x}$ is the sample mean. Assuming a normal distribution, test, at the 5% significance level, whether the allegation appears reasonable.

2 A sample of eight boxes of soap powder is selected at random from a large batch. The contents of the boxes have a nominal weight of 2000 grams. After subtracting 2000 grams the actual weights are:

$-1.5 \quad 0.4 \quad -0.1 \quad 5.8 \quad 11.5 \quad 7.6 \quad 1.3 \quad 2.4$

Assuming a normal distribution for the weights of the contents, show that there is significant evidence, at the 5% level, that the mean weight of the contents of the containers in this batch is greater than 2000 grams.

3 A machine tests reaction times. A random sample of ten adults record the following times (in seconds):

$5.2 \quad 5.5 \quad 4.9 \quad 5.4 \quad 4.9 \quad 5.5 \quad 4.9 \quad 5.8 \quad 4.9 \quad 5.0$

Making suitable assumptions, which should be stated, test, at the 5% level, the hypothesis that the average reaction time is 5.0 seconds.

4 The lifetimes of Shinelong bulbs are normally distributed. It is claimed that this distribution has a mean in excess of 1200 hours. To test this assertion, a random sample of 12 bulbs will be taken. The actual lifetimes will be recorded in the form $(1200 + x)$ hours, where x is the excess lifetime (which will be negative if a bulb fails before 1200 hours). A t-test will be carried out, using a 5% significance level.

a) Find the critical region for t.

b) State, in the context of the question, what is meant by a 'Type I error', and state the probability of such an error occurring.

c) State, in the context of the question, what is meant by a 'Type II error'.

d) Given that the sample results are summarised by $\sum x = 93.4$, $\sum(x - \bar{x})^2 = 6453.62$, where $\bar{x}$ is the sample mean excess, carry out the test.

e) Suppose instead that $\sum x = -93.4$ and that the value of $\sum(x - \bar{x})^2$ has been mislaid. Explain why the outcome of the t-test is nevertheless known.

5 It is claimed that a new variety of fruit bush gives a higher yield than the variety it will replace. A random sample of eight bushes of the new variety is grown and the yield of each bush is recorded. The old variety had an average yield of 4.7 kg per bush. The yields, x kg, of the sample of new bushes are summarised by $\sum x = 52.4$, $\sum(x - \bar{x})^2 = 19.13$, where $\bar{x}$ is the sample mean. It may be assumed that each item of data is an independent observation from a normal distribution. Test the claim at the 1% level.

6 A jam manufacturer produces thousands of jars of strawberry jam each week. The weight of jam in a jar is an observation from a normal distribution having mean 455 grams and standard deviation 0.9 grams. Determine the probability that a randomly chosen jar contains less than 453 grams.

Following a slight adjustment to the filling machine, a random sample of 9 jars is found to contain the following weights (in grams) of jam:

454.7 453.9 455.1 454.2 455.3
454.5 454.4 455.1 449.9

a) Assuming that the variance of the distribution is unaltered by the adjustment, test, at the 5% significance level, the hypothesis that there has been no change in the mean of the distribution.

b) Assuming that the variance of the distribution may have altered, obtain an unbiased estimate of the new variance and, using this estimate, test, at the 5% significance level, the hypothesis that there has been no change in the mean of the distribution.

S2

5.5 Test for the mean of a distribution using a normal approximation

Suppose a sample is taken from a population with known variance σ^2. The distribution of the population is either unknown, or is known not to be normal. The tests of the previous sections relied upon the population distribution being normal, so these appear to be inappropriate.

However, if the sample size is large ($n \geqslant 30$, say) then, as a consequence of the Central Limit Theorem, the random variable Z, given by:

$$Z = \frac{\overline{X} - \mu_0}{\sqrt{\frac{\sigma^2}{n}}}$$

has an approximate N(0, 1) distribution. Here $\overline{X}$ is the random variable corresponding to the sample mean and μ_0 is the mean specified by a null hypothesis. Thus a z-test is once more appropriate.

The Central Limit Theorem was covered in Chapter 5 of *Statistics S1*.

Often the variance will also be unknown. With a sample size in excess of 30, the unbiased estimate, s^2, will be a reasonably accurate estimate. Substituting s^2 for σ^2 in the previous statistic gives:

$$\frac{\overline{X} - \mu_0}{\sqrt{\frac{s^2}{n}}}$$

Since s^2 is involved, it might be thought that a t_{n-1}-distribution would be appropriate, but, strictly, that requires a normal population. In this case the distribution of $\overline{X}$ is only approximately normal. For this reason, despite the use of s^2, it is expected that the critical values of an N(0, 1) distribution will be used.

In an examination a candidate would not usually be penalised for using critical values from a t_{n-1}-distribution or, equivalently, carrying out a t-test on a calculator.

Example 7

In an experiment on people's perception, a class of 100 students were given a piece of paper which was blank except for a line 120 mm long. The students were asked to judge by eye the centre point of the line, and to mark it. The students then measured the distance, x mm, between the left-hand end of the line and their mark. Working with $y = x - 60$, the results are summarised by $\sum y = -143.5$, $\sum y^2 = 1204.00$.

Determine whether there is evidence, at the 1% significance level, of any overall bias in the students' perception of the centre of the lines.

1. *Write down H_0 and H_1.*
   The test is two tailed since there is no suggestion in the question that any bias will necessarily be to the left or to the right. Working with Y (with mean μ) the hypotheses are therefore

   H_0: $\mu = 0$
   H_1: $\mu \neq 0$

2. *Identify an appropriate test statistic and the distribution of the corresponding random variable (using the parameter value specified by H_0).*
   Since σ^2 is unknown, but n is large, s^2 is used. The test statistic is therefore:

   $$z = \frac{\bar{y} - 0}{\sqrt{\frac{s^2}{100}}}$$

   which, assuming H_0, will be an observation from an approximate standard normal distribution.

3. *Identify the significance level.*
   The question specifies 1%.

4. *Determine the critical region.*
   The test is two tailed. Since $P(Z > 2.5758) = 0.005$, and $P(Z < -2.5758) = 0.005$, an appropriate procedure is to accept H_0 if z lies in the interval $(-2.5758, 2.5758)$ and otherwise to reject H_0 in favour of H_1. Thus the critical region is given by $|z| > 2.5758$.

5. *Calculate the value of the test statistic.*
   Since $\bar{y} = \frac{-143.5}{100} = -1.435$ and s^2 is given by:

   $$s^2 = \frac{1}{99}\left(1204.00 - \frac{(-143.5)^2}{100}\right) = 10.0816$$

S2

the test statistic is given by:

$$z = \frac{-1.435 - 0}{\sqrt{\frac{10.0816}{100}}} = -4.52$$

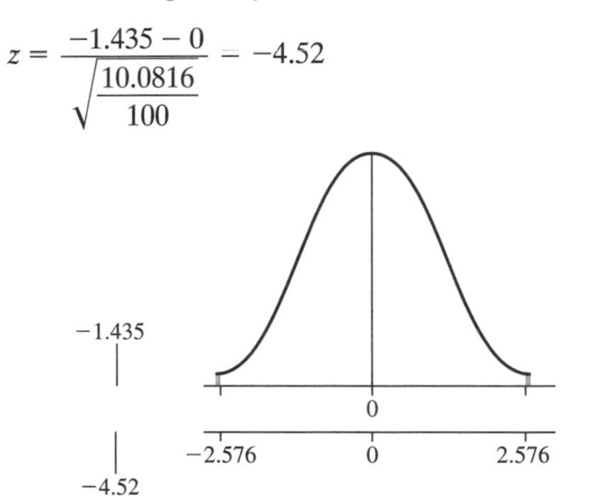

6. *Determine the outcome of the test.*
   Since $z < -2.5758$, H_0 is rejected and thus H_1 is accepted. There is evidence, at the 1% level, that the students' results are biased. Indeed, the result would also have been deemed significant at the 0.001% level!

Example 8

The lengths (x kilobytes) of a random sample of 1000 messages from the inbox of a busy company are summarised by $\sum x = 17\,405$ and $\sum(x - \bar{x})^2 = 3243$, where $\bar{x}$ is the sample mean. Test, at the 0.1% significance level, the hypothesis that the mean length of the messages in the inbox exceeds 17 kilobytes.

1. *Write down H_0 and H_1.*
   The hypotheses are:

   H_0: $\mu = 17$
   H_1: $\mu > 17$

2. *Identify an appropriate test statistic and the distribution of the corresponding random variable (using the parameter value specified by H_0).*
   Since σ^2 is unknown, but n is large, s^2 is used and the test statistic is therefore:

   $$z = \frac{\bar{x} - 17}{\sqrt{\frac{s^2}{1000}}}$$

   which, assuming H_0, will be an observation from an approximate standard normal distribution.

> To carry out this test on a calculator you will need to input the value of s as the value of σ.

3. *Identify the significance level.*
   The question specifies 0.1%.

4. *Determine the critical region.*
   The test is one tailed.

   From Table 4 in the Appendices, the critical value for Z is 3.0902, since $P(Z > 3.0902) = 0.001$. The procedure is to accept H_0 if $z < 3.0902$ and otherwise to reject H_0 in favour of H_1. Thus the critical region is given by $z > 3.0902$.

5. *Calculate the value of the test statistic.*
   Since $\bar{x} = 17.405$, and s^2 is given by

   $$s^2 = \frac{3243}{999} = 3.2462$$

   the test statistic is:

   $$z = \frac{17.405 - 17}{\sqrt{\frac{3.2462}{1000}}} = 7.11$$

6. *Determine the outcome of the test.*
   Since $z > 3.0902$, H_0 is rejected and thus H_1 is accepted.

   There is very strong evidence, at the 0.1% level, that the average length of a message in the inbox exceeds 17 kilobytes.

S2

Exercise 5E

1 A teacher notes the time that she takes to drive to school. She finds that, over a long period, the mean time is 24.5 minutes. After a new bypass is opened, she notes the time on each of 72 randomly chosen journeys to school. The mean of these times is 23.0 minutes, with an unbiased estimate of variance being 45.66 minutes^2. Using a 5% significance level, test whether the journey now takes less time.

2 The mean IQ score is adjusted to be 100 for each age group of the population. A random sample of 3-year-old children is given vitamin supplements for five years. At the end of the period the 180 children have mean IQ score 102.4. The sample variance is 219.4. Test whether there is significant evidence at the 1% level to support the theory that vitamin supplements increase IQ scores.

3 An inspector wishes to determine whether eggs sold as Size 1 have mean weight 70.0 g or not. She intends to carry out a test, at the 1% significance level.

a) State, in the context of the question, what is meant by:
   i) a Type I error ii) a Type II error.

b) The inspector weighs a sample of 200 eggs and her results are summarised by $\sum x = 13\,824$ and $\sum (x - \bar{x})^2 = 3320$, where x is the weight of an egg in grams and $\bar{x}$ is the sample mean. Carry out the test.

4 Rumour has it that the average length of a leading article in the 'Daily Intellectual' is 960 words. As part of a project, a student counts the number of words in each of 55 randomly chosen leading articles from the paper. His results give $\sum x = 53\,392$ and $\sum(x - \bar{x})^2 = 146\,729$, where $\bar{x}$ denotes the sample mean. Test, at the 10% significance level, the truth of the rumour.

5 A supermarket manager investigated the lengths of time that customers spent shopping in the store. The time, x minutes, spent by each of a random sample of 160 customers was measured and it was found that $\sum x = 3092$ and $\sum(x - \bar{x})^2 = 1517$, where $\bar{x}$ is the sample mean. Test, at the 5% level of significance, the hypothesis that the mean time spent shopping by customers is 20 minutes, against the alternative that it is less than this.

S2

6 The heights of 100 individuals, randomly sampled from a population of school-children, are summarised in the following table.

Height range (cm)	120–	130–	140–	150–	160–	170–190
Frequency	5	17	20	30	22	6

Test, at the 5% significance level, the hypothesis that the mean height of the school-children is 1.5 m.

Hypothesis tests and confidence intervals

There is a simple rule that usually works in the case of a two tailed alternative hypothesis:

> *If an α% confidence interval excludes the population value of interest, then the null hypothesis that the population parameter takes this value will be rejected at the* $(100 - \alpha)$% *level.*

Suppose, for example, that the 95% confidence interval for a population mean, μ, is (83.0, 85.1). In a two tailed test, the null hypothesis that $\mu = 85.2$ will be rejected at the 5% level since the interval excludes 85.2. Indeed *any* hypothesised value for μ that is greater than 85.1 or is less than 83.0, will be rejected at the 5% level. Conversely, the hypothesis that μ takes a specific value in the range (83.0, 85.1) will be accepted at the 5% level.

Example 9

A machine cuts wood to form stakes, which are supposed to be 2 m long. A random sample of 40 stakes are accurately measured, and their lengths, $(200 + x)$ cm, are summarised by:

$$\sum x = 8041.6 \quad \text{and} \quad \sum(x - \bar{x})^2 = 67.46$$

Determine a 95% confidence interval for the mean stake length, giving the limits to two decimal places. Perform a two tailed test, at the 5% significance level, of the null hypothesis that the population mean is 2 m.

The unbiased estimates of the population mean and variance are given by:

$$\bar{x} = \frac{8041.6}{40} = 201.04 \quad \text{and} \quad s^2 = \frac{67.46}{39} = 1.7297$$

The sample size is sufficiently large that the distribution of the sample mean can be taken to be normal (because of the Central Limit Theorem). There will be little loss of accuracy in treating $\dfrac{\bar{X} - \mu}{\frac{s}{\sqrt{n}}}$ as having N(0, 1) distribution. The 95% confidence interval is therefore:

$$201.04 \pm 1.9600\sqrt{\frac{1.7297}{40}} = (200.63, 201.45)$$

Since the interval excludes 200, the hypothesis that the mean is 2 m is rejected, at the 5% significance level, in favour of the alternative that this is not the case.

For a sample from a normal distribution, the confidence interval would use critical values from the appropriate *t*-distribution.

S2

Exercise 5F

1 Jars of honey are filled by a machine. It has been found that the quantity of honey in a jar has mean 460.3 grams, with standard deviation 3.2 grams. It is believed that the controls have been altered in such a way that, although the standard deviation is unaltered, the mean quantity may have changed. A random sample of 60 jars is taken and the mean quantity of honey per jar is found to be 461.2 grams. State suitable null and alternative hypotheses, and carry out a test using a 5% level of significance, by finding an appropriate confidence interval.

2 Observations of the time taken to test an electrical circuit board show that it has mean 5.82 minutes with standard deviation 0.63 minutes. As a result of the introduction of an incentive scheme, it is believed that the inspectors may be carrying out the test more quickly. It is found that, for a random sample of 150 tests, the mean time taken is 5.68 minutes. State suitable null and alternative hypotheses. Assuming that the population variance remains unchanged, carry out a test at the 5% significance level, by finding an appropriate confidence interval.

3 A light bulb manufacturer has established that the life of a bulb has mean 95.2 days with standard deviation 10.4 days. Following a change in the manufacturing process, which is not intended to change the mean life of a bulb, a random sample of 96 bulbs has mean life 93.6 days. State suitable hypotheses. Assuming that the population standard deviation is unchanged, test whether there is significant evidence, at the 1% level, of a change in mean life, by finding an appropriate confidence interval.

Here is a table that summarizes the terminology introduced in this chapter.

Terminology

Null hypothesis (H_0)	An equation specifying that a parameter in a statistical model takes a particular value. The test of the null hypothesis is based on the assumption that the parameter has that value.
Alternative hypothesis (H_1)	This expresses the way in which a parameter may deviate from the value specified by H_0. This hypothesis is accepted when H_0 is rejected.
Type I error	Rejection of H_0 when it is, in fact, true.
Type II error	Acceptance of H_0 when it is, in fact, false.
Test statistic	A function of a sample of observations which provides a basis for testing the validity of the null hypothesis.
Critical region	The null hypothesis is rejected when the value of the test statistic lies in this region.
Critical value	A value determining a boundary of the critical region.
Significance level ($\alpha\%$)	The 'size' of the critical region. It is the probability of a Type I error.
One tailed test	A test appropriate when H_1 involves $>$ or $<$.
Two tailed test	A test appropriate when H_1 involves $\neq$.

Summary

You should now be able to ...	Check out
1 Understand and use the language of hypothesis tests.	**1** A random variable has a $N(\mu, 100)$ distribution. A 5% significance test of whether μ has increased from its current value of 250 shows $z = \dfrac{255 - 250}{\sqrt{\dfrac{100}{16}}} = 2.$ a) Write down the null and alternative hypotheses. b) State whether the test is one tailed or two tailed. c) Identify the test statistic. d) Find the critical region and critical value. e) Explain, in context, the meaning of Type I and Type II errors. f) State the conclusion of the test.
2 Test for the mean of a normal population with known variance.	**2** A random variable, X, has a $N(\mu, 25)$ distribution. Test, at the 10% significance level, H_0: $\mu = 80$, against H_1: $\mu \neq 80$, given that a random sample of size 9 has a mean of 78.
3 Test for the mean of a normal population with unknown variance.	**3** A random variable, X, has a $N(\mu, \sigma^2)$ distribution A random sample of size 16 gives $\bar{x} = 21.2$ and $s^2 = 2.56$. Investigate, at the 1% significance level, the claim that $\mu > 20$.
4 Test for the mean of a population using a normal approximation.	**4** The random variable, Y, has a distribution with mean μ and variance σ^2. A random sample of size 50 gives $\bar{y} = 252.6$ and $s^2 = 130.24$. Test, at the 5% significance level, the claim that $\mu = 250$, with the alternative hypothesis being that $\mu \neq 250$.
5 Identify the relationship between confidence intervals and hypothesis tests.	**5** A 90% confidence interval for μ is (240.2, 244.8). Use this to test, at the 10% significance level, H_0: $\mu = 243$ against H_1: $\mu \neq 243$.

Revision exercise 5

1 The lengths of fish of a particular species are normally distributed with a mean of 56 cm and a standard deviation of 4.2 cm. There is a suspicion that, due to overfishing, the mean length of these fish has changed.

A random sample of 50 fish was measured and was found to have a mean length of 54.8 cm.

Investigate, at the 5% level of significance, whether this indicates a change from 56 cm in the mean length of this species of fish. *(AQA, 2001)*

2 A random variable X is normally distributed with mean μ and standard deviation 0.8. The null hypothesis H_0: $\mu = 40$ is to be tested against the alternative hypothesis H_1: $\mu \neq 40$ using the 5% level of significance.

The mean, $\overline{X}$, of a random sample of 50 observations of X is to be used as the test statistic.

a) Write down:
   i) the distribution of $\overline{X}$ assuming H_0 is true,
   ii) the probability of a Type I error.

b) Calculate the acceptance region for $\overline{X}$, giving its limits to two decimal places.

c) Explain what is meant by a Type II error. *(AQA, 2002)*

3 The volume, Y ml, of washing-up liquid contained in a plastic bottle is a normal random variable with mean μ and standard deviation 8.

It is suspected that μ has a value which is less than the specified value of 500.

To test this suspicion, a random sample of 100 bottles of washing-up liquid is selected and the mean volume, $\overline{Y}$, of liquid per bottle is calculated.

a) Write down null and alternative hypotheses for this test.

b) State the distribution of $\overline{Y}$, assuming the null hypothesis to be true.

c) Assuming a 5% significance level:
   i) write down the probability of a Type I error,
   ii) determine the critical region for $\overline{Y}$, giving its limit to two decimal places. *(AQA, 2003)*

4 Bill is an athlete specialising in the long jump. It is known that the distance he jumps can be regarded as a normally distributed random variable with standard deviation 0.1 m.

Bill claims that the mean length of his jumps is 6 m. To test this claim, he jumps six times and the distances jumped, in metres, are as follows.

5.78 6.09 5.86 5.92 6.01 5.74

Investigate Bill's claim using a two tailed test with significance level 10%. *(AQA, 2002)*

5 A woodland area in France is home to a population of dormice. The weights of adult male dormice in this population are normally distributed.

A conservationist takes a random sample of 8 adult dormice in June and records the weight, x grams, of each dormouse. He calculates the sample mean, $\bar{x}$, and an unbiased estimate, s^2, of the population variance with the following results.

$\bar{x} = 17.51 \quad s^2 = 1.273$

Test, at the 10% significance level, the claim that the mean weight of adult male dormice in this population in June is 17 grams. (*AQA, 2004*)

6 The mean payment made by an insurance company on household claims in 1999 was £632. The company's payments, in £, on a sample of claims in 2000 were

459 234 2067 198 342 456 732 219 701

a) Stating clearly your null and alternative hypotheses, investigate, at the 5% significance level, whether there has been a change in the mean payment in 2000 compared to that in 1999.

b) State **two** assumptions you had to make when carrying out the test in part a).

The manager now examined 180 randomly selected claims made in 2000. She found the mean payment was £768 with a standard deviation of £299.

c) Stating clearly your null and alternative hypotheses, use the results from these 180 claims to investigate, at the 1% significance level, whether the mean payment has increased in 2000 compared to that in 1999.

d) Explain why you do not need to make the assumptions you stated in part b) in order to carry out the test in part c).

e) If, in part a), you had been asked to investigate whether the mean payment had increased in 2000 when compared to that in 1999, explain why a calculation of the sample mean would be sufficient to enable you to answer the question. (*AQA, 2001*)

7 An instrument for measuring the speed of passing motorists is tested by a police force. A car is driven at known speeds down a straight road. The error (speed recorded by the instrument minus speed of the car) is observed on eight occasions with the following results in m/s:

4.2 −2.8 3.7 −5.9 0.2 6.4 4.1 −1.9

a) Stating your null and alternative hypotheses, investigate, at the 5% significance level, whether the instrument is biased (i.e. whether the mean error differs from zero). Assume that the data may be regarded as a random sample from a normal distribution.

b) Tests on a different design of instrument showed that in 20 trials there was a mean error of −0.25. From past experience it is known that these errors are normally distributed with a standard deviation of 0.38.

i) Investigate, using the 1% significance level, whether this second instrument, on average, underestimates the speed of cars.

ii) Compare the performance of the two instruments and give a reason why the second instrument might be preferred to the first. *(AQA, 2002)*

8 A publisher produced a quick revision book for A-Level Mathematics. To demonstrate its effectiveness, the book was offered free of charge to a class of students, two weeks before they took a mock A-Level examination.
The marks, in the mock examination, of eight of the students who accepted the offer were as follows:

73 49 64 46 73 55 59 52

The mean mark in the mock examination, based on past experience, was expected to be 50.

a) Investigate, at the 5% significance level, whether the mean mark for students who accepted the offer differed from 50. State your null and alternative hypotheses.

b) State the assumptions you have made in order to carry out the test in part a).

c) Encouraged by the initial trial the publisher offered the book free of charge to all students in a particular local education authority area who were due to take A-level Mathematics.

A sample of 89 students, who accepted the offer and used the book, obtained a mean mark of 50.4 with a standard deviation of 12.6 in the examination. The average mark for all students who took the same A-level Mathematics examination without using the revision book was 48.8.

i) Stating your null and alternative hypotheses, examine, at the 1% significance level, whether the mean mark for students who used the book was greater than 48.8. Assume that the data satisfies all conditions necessary for carrying out the test.

ii) Give a reason (other than the possibility of making a Type I error) why, however high the marks obtained by the students who used the book, this further trial cannot prove that the book is an effective way of improving A-Level Mathematics marks. *(AQA, 2002)*

9 A make of car battery is advertised as having a mean lifetime of at least 40 months. A car magazine suspects that the mean lifetime is less than 40 months and so monitors the lifetimes, in months, of a random sample of 12 batteries, with the following results:

44 32 46 39 38 30 35 40 38 41 34 36

a) Assuming that lifetimes are normally distributed, investigate the magazine's suspicion, using the 5% significance level.

b) A larger random sample of 160 car batteries is now monitored. The lifetimes are found to have a mean of 39.2 months and a standard deviation of 4.2 months. Use this information to investigate the magazine's suspicion, using the 5% significance level.

c) State, with an explanation, which, if either, of the tests in parts a) and b) has the larger risk of a Type I error. *(AQA, 2004)*

6 Chi-squared (χ^2) contingency table tests

This chapter will show you how to

- Use tables of the chi-squared distribution
- Calculate an approximate chi-squared statistic
- Test the hypothesis of independence for an $r \times c$ contingency table

Before you start

S2

You should know how to ...	Check in
1 Apply the multiplication law of probability to two independent events.	**1** An experiment consists of tossing a fair coin and rolling a fair six-sided die. Calculate the probability of obtaining: a) a head and a six b) a tail and (a one or a two).
2 Determine the expected frequency of an event.	**2** For the example in question **1** a), calculate expected frequencies when the experiment is repeated: a) 60 times b) 100 times.

6.1 Introduction to the χ^2 distribution

'Chi' is the Greek letter χ, pronounced 'kye'. The chi-squared distribution is continuous and has a positive integer parameter ν, which determines its shape. The distribution has a number of uses in Statistics. One of these is described later in this chapter.

Recall that ν is the Greek letter 'nu'.

As in the case of the t-distribution, the parameter ν is known as the **number of degrees of freedom** of the distribution and the distribution is referred to as a 'chi-squared distribution with ν degrees of freedom'. For simplicity, this is written as:

$$\chi^2_\nu$$

The probability density function of χ^2 describes the distribution of a non-negative random variable (as suggested by the 'squared' in the distribution's name).

In fact, if Z has an N(0,1) distribution, then Z^2 has a χ^2_1 distribution.

The figure shows parts of the graphs of the probability density functions of chi-squared distributions having 1, 2, 4, 8 and 16 degrees of freedom. In each case the minimum value of the variable is 0 and there is no maximum value.

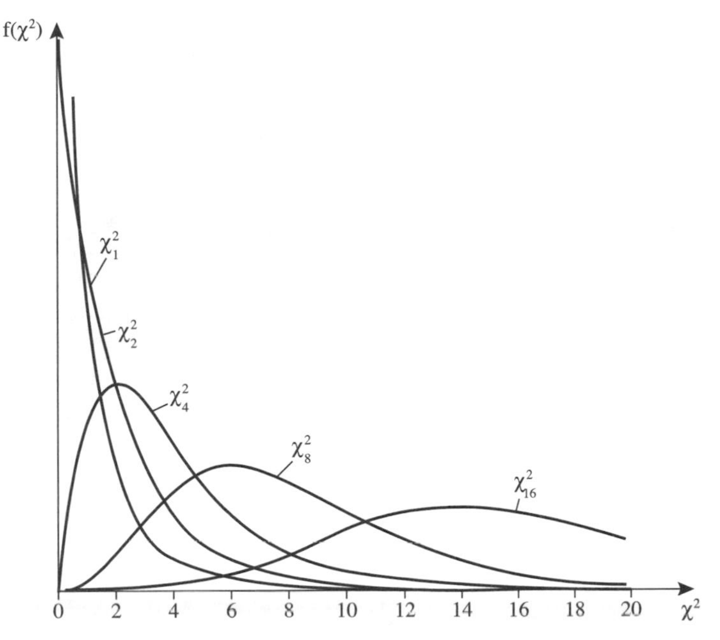

A χ^2_ν distribution has mean ν and variance 2ν.

For $\nu > 2$ the probability density function has a maximum at $\nu - 2$.

As the figure suggests, the distribution is always skewed (non-symmetric), with the skewness being most marked for low values of ν.

Tables of the chi-squared distribution

Table 6 is reproduced in the Appendices.

Tables for the chi-squared distribution concentrate on percentage points. Here is an extract from Table 6 of the AQA tables.

p	**0.005**	**0.01**	**0.025**	**0.05**	**0.1**	**0.9**	**0.95**	**0.975**	**0.99**	**0.995**	p
ν											ν
1	0.00004	0.0002	0.001	0.004	0.016	2.706	3.841	5.024	6.635	7.879	**1**
2	0.010	0.020	0.051	0.103	0.211	4.605	5.991	7.378	9.210	10.597	**2**
3	0.072	0.115	0.216	0.352	0.584	6.251	7.815	9.348	11.345	12.838	**3**
4	0.207	0.297	0.484	0.711	1.064	7.779	9.488	11.143	13.277	14.860	**4**
5	0.412	0.554	0.831	1.145	1.610	9.236	11.070	12.833	15.086	16.750	**5**
⋮	⋮	⋮	⋮	⋮	⋮	⋮	⋮	⋮	⋮	⋮	⋮
39	19.996	21.426	23.654	25.695	28.196	50.660	54.572	58.120	62.428	65.476	**39**
40	20.707	22.164	24.433	26.509	29.051	51.805	55.758	59.342	63.691	66.766	**40**
45	24.311	25.901	28.366	30.612	33.350	57.505	61.656	65.410	69.957	73.166	**45**
⋮	⋮	⋮	⋮	⋮	⋮	⋮	⋮	⋮	⋮	⋮	⋮
100	67.328	70.065	74.222	77.929	82.358	118.498	124.342	129.561	135.807	140.169	**100**

Table 6 contains two sections, one dealing with the lower tail, and one with the upper tail. The percentage points, for various values of the cumulative probability f, are given for every value of ν between 1 and 40, and then in steps of 5 to 100, as indicated.

Thus, reading from the table:

$$P(\chi_5^2 < 1.610) = 0.1$$
$$P(\chi_5^2 > 9.236) = 1 - 0.9 = 0.1$$

The table also enables the determination of the value of a chi-squared random variable that would be exceeded with some specified probability. For example, the upper 1% point of a χ_3^2 distribution is 11.345.

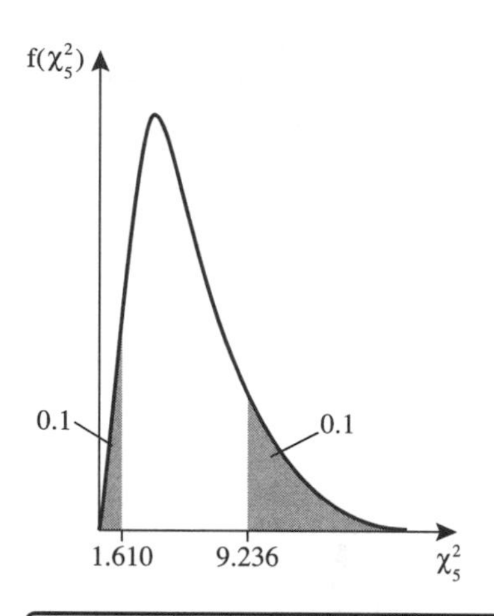

Because the distribution is not symmetric, there is no easy relationship between, for example, the upper 10% point and the lower 10% point.

S2

Exercise 6A

1 Find:

a) $P(\chi_1^2 < 3.841)$ b) $P(\chi_2^2 < 9.210)$ c) $P(\chi_{39}^2 < 62.428)$.

2 Find:

a) $P(\chi_4^2 > 11.143)$ b) $P(\chi_{40}^2 > 63.691)$ c) $P(\chi_{45}^2 > 57.505)$.

3 Find c such that:

a) $P(\chi_4^2 > c) = 0.005$ b) $P(\chi_1^2 > c) = 0.05$
c) $P(\chi_{40}^2 < c) = 0.95$ d) $P(\chi_2^2 < c) = 0.995$.

4 Find:

a) $P(7.815 < \chi_3^2 < 12.838)$ b) $P(57.505 < \chi_{45}^2 > 61.656)$.

6.2 Use of $\sum \frac{(O_i - E_i)^2}{E_i}$ as an approximate χ^2 statistic

The section heading will be explained a little later in the section, after introducing the context in which the statistic is used.

To motivate matters, suppose that a tetrahedral die with its four faces painted in different colours (red, white, blue and green) has been tossed forty times, with the following results:

Outcome	Red	White	Blue	Green
Observed frequency	18	1	12	9

A tetrahedral die is a pyramid; the face chosen is the bottom face.

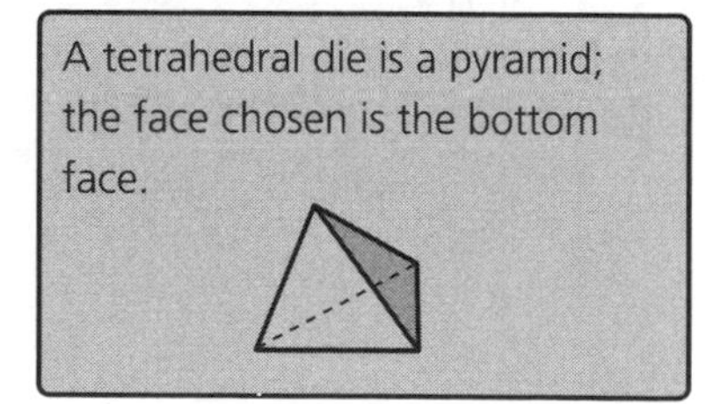

The question of interest is whether the die is biased. If the die were fair then each face would be equally likely. With 40 tosses the expected frequencies would be $40 \times \frac{1}{4} = 10$:

Outcome	Red	White	Blue	Green
Expected frequency	10	10	10	10

Obviously to judge whether the observed frequencies (O) differ from the expected frequencies (E), it is necessary to look at the differences in their values ($O - E$):

Outcome	Red	White	Blue	Green
Observed frequency, O	18	1	12	9
Expected frequency, E	10	10	10	10
Difference, $O - E$	8	−9	2	−1

The larger the magnitude of the differences, the more the observed data differs from that expected according to the model.

Suppose a second die is rolled 440 times with the following results:

Outcome	Red	White	Blue	Green
Observed frequency, O	118	101	112	109
Expected frequency, E	110	110	110	110
Difference, $O - E$	8	−9	2	−1

This time the observed and expected frequencies seem remarkably close, yet the $O - E$ values are the same as before. Evidently it is not simply the size of $O - E$ that matters, but also its size relative to the expected frequency: $\frac{O - E}{E}$.

Combining the ideas that both 'difference' and 'relative size' matter suggests using the product $(O - E) \times \frac{O - E}{E}$, so that the goodness of fit for outcome i may be measured using $\frac{(O_i - E_i)^2}{E_i}$. The smaller this quantity is, the better the fit. An aggregate measure of **goodness of fit** of the model is therefore provided by X^2, defined by:

$$X^2 = \sum_{i=1}^{m} \frac{(O_i - E_i)^2}{E_i} \qquad (6.1)$$

where m is the number of different outcomes (4, in the case of the die). Significantly large values of X^2 suggest lack of fit.

A useful check that you should carry out before starting on the calculation of X^2 is to make sure that the total of your expected frequencies is equal to the total of the observed frequencies. A convenient way of doing this is provided by checking that your $(O_i - E_i)$ values sum to zero.

The O_i are observed frequencies (*not percentages or measurements*) and are always whole numbers. However, the E_i will not usually be whole numbers and they should be calculated accurately.

Different samples (for example, different sets of 40 rolls of the die) will give different sets of observed frequencies and hence different values for X^2. Thus X^2 has a probability distribution. The exact distribution is difficult to compute, but it is usually well approximated by a chi-squared distribution. For a case where the probabilities of the m possible outcomes are completely prescribed, the distribution of X^2 is approximately χ^2_{m-1}.

Example 1

For the tetrahedral die described on p 144, test, at the 5% level, the null hypothesis that the four faces are equally likely:

a) using the given data for the case of 40 rolls

b) using the given data for the case of 440 rolls.

The null hypothesis, H_0, is that the four faces are equally likely, with the alternative hypothesis, H_1, being that this not the case. For this simple null hypothesis the expected frequencies are easily calculated by dividing the number of rolls by 4.

a) It is a good idea to set the calculations neatly in a table:

Face	O_i	E_i	$O_i - E_i$	$\frac{(O_i - E_i)^2}{E_i}$
Red	18	10	8	6.4
White	1	10	−9	8.1
Blue	12	10	2	0.4
Green	9	10	−1	0.1
Total	40	40	0	15.0

How has the large value of X^2 (15.0) come about? Studying the final column of the table, it is apparent that the major contributions come from the first two rows. On the basis of this set of data, the die appears to be biased towards Red and away from White.

In this case $m = 4$ and the relevant χ^2 distribution therefore has 3 degrees of freedom. From Table 6, the upper 5% point of a χ_3^2 distribution is 7.815, which is less than the observed value of X^2 (15.0). Thus, at the 5% level, the null hypothesis that the four faces are equally likely is rejected.

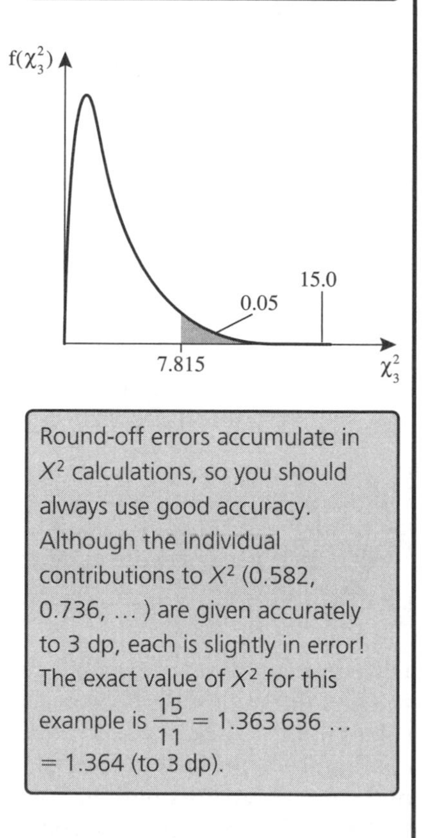

b) The calculations follow the same pattern:

Face	O_i	E_i	$O_i - E_i$	$\frac{(O_i - E_i)^2}{E_i}$
Red	118	110	8	0.582
White	101	110	−9	0.736
Blue	112	110	2	0.036
Green	109	110	−1	0.009
Total	440	440	0	1.363

The $O_i - E_i$ values are the same as in part a), but the expected frequencies are much greater. The result is a much smaller value for X^2.

Since 1.363 is much smaller than 7.815, at the 5% level the null hypothesis that the four faces are equally likely is accepted.

Note that since the calculated value of X^2 (1.363) is less than the mean of the approximating χ_3^2 distribution (which has mean 3), the observed value cannot lie in the upper tail of the distribution.

Round-off errors accumulate in X^2 calculations, so you should always use good accuracy. Although the individual contributions to X^2 (0.582, 0.736, ...) are given accurately to 3 dp, each is slightly in error! The exact value of X^2 for this example is $\frac{15}{11} = 1.363\,636\ldots$ $= 1.364$ (to 3 dp).

6.3 Conditions for approximation to be valid

The distribution of X^2 is discrete, whereas the approximating χ^2 distribution is continuous. The approximation becomes less accurate as the expected frequencies become smaller. The convention used for the module Statistics S2 is that:

> *'All expected frequencies must be greater than 5'.*

If one of the originally chosen categories has an expected frequency of 5 or less, then this category must be combined with another category. This combination may be done on any sensible grounds, but should be done *without reference to the observed frequencies* so as to avoid biasing the results.

- With numerical data it is natural to combine numerically adjacent categories. For example, the three categories '7', '8' and '9' might be replaced by the single category '7–9'.
- With categorical data it is natural to combine categories that are related. For example, in comparing the frequencies of birds seen in farmland against birds seen in town, it might be sensible to combine the separate counts for greenfinches, chaffinches, and so on, into a single category 'finches', with the counts for jackdaws, rooks, and so on being combined into a single category 'crow family'.

Example 2

A six-sided die has one face marked '1', one face marked '2', two faces marked '3' and two faces marked '4'. The die is rolled 24 times with the following results:

Face	1	2	3	4
Frequency	8	3	11	2

Use a 5% significance level to decide whether these results suggest that the die is biased.

The null hypothesis (H_0) is that the die is unbiased, which would imply that each of the six faces is equally likely (the alternative hypothesis, H_1, is that the die is biased). With 24 rolls, each face is therefore expected on $\frac{24}{6} = 4$ occasions. Thus the expected frequencies for the outcomes '1', '2', '3' and '4' are 4, 4, 8 and 8, respectively.

Two expected frequencies are less than 5, so each of the corresponding categories must be combined with some other category. A natural choice (without reference to the observed frequencies) is to combine these two categories with each other.

The resulting calculations are then as follows:

Face	O_i	E_i	$O_i - E_i$	$\frac{(O_i - E_i)^2}{E_i}$
1 or 2	$8 + 3 = 11$	$4 + 4 = 8$	3	1.125
3	11	8	3	1.125
4	2	8	−6	4.500
Total	24	24	0	6.750

Despite its small observed frequency, category '4' is not combined with another category. This is because its expected frequency exceeds 5.

After combining categories, three categories remain. There are therefore $(3 - 1) = 2$ degrees of freedom for the test statistic. Since the observed value of X^2 (6.750) exceeds the upper 5% critical value of a χ_2^2 distribution (5.991, see Table 6), there is evidence, at the 5% level, to reject H_0.

S2

There is significant evidence that the die is biased.

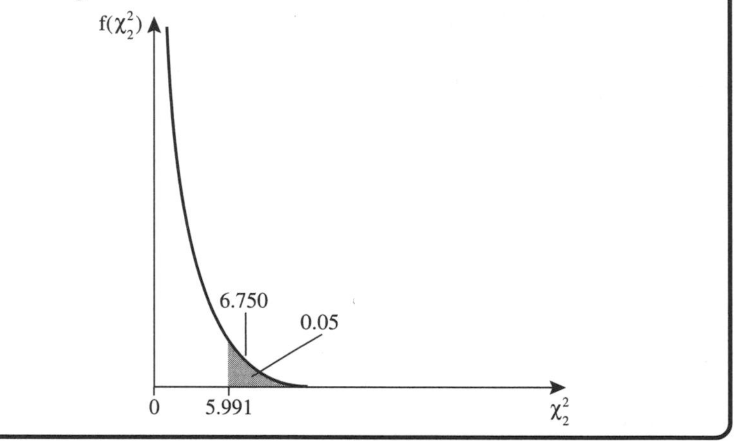

6.4 Tests for independence in contingency tables

Often data are collected on several variables at a time. For example, a questionnaire will usually contain more than one question! A table that gives the frequencies for two or more variables simultaneously is called a **contingency table**. Here is an example which shows information on voting:

	Conservative	Liberal Democrat	Labour	
Male	313	124	391	828
Female	344	158	388	890
	657	282	779	1718

Sample data of this type are collected in order to answer interesting questions about the behaviour of the population. such as 'Are there differences in the way that males and females vote?'. If there are differences then the variables *vote* and *gender* are said to be **associated**, whereas if there are no differences then the variables are said to be **independent**.

The null hypothesis is that the variables are independent. If this is true then, in the population, the proportion of Conservative supporters who are male will be equal to the proportion of Liberal Democrat supporters who are male and to the proportion of Labour supporters who are male. Furthermore, the proportion of males who support the Conservative party will be equal to the proportion of females who support the Conservative party, and so on.

The alternative hypothesis is that the variables are not independent.

The best estimate of the proportion of males in the population is $\frac{828}{1718}$. Similarly, the best estimate of the population proportion voting for the Conservatives is $\frac{657}{1718}$. According to the null hypothesis these characteristics are independent, so that the best estimate of the proportion of the population who are males voting for the Conservatives is:

$$\frac{828}{1718} \times \frac{657}{1718}$$

and thus the expected number of males voting for the Conservatives is:

$$1718 \times \frac{828}{1718} \times \frac{657}{1718} = \frac{828 \times 657}{1718} = 316.645 \text{ (to 3 dp)}$$

S2

The expected value for any cell in a contingency table may be calculated using the formula:

$$\frac{\text{Row total} \times \text{Column total}}{\text{Grand total}}. \qquad (6.2)$$

The full set of expected frequencies (to 3 dp) are:

316.645	135.912	375.444	828.001
340.355	146.088	403.556	889.999
657.000	282.000	779.000	

Note that the row totals and column totals of the expected frequencies should always be identical to those for the observed frequencies. However rounding can lead to inaccuracies (as here for the row totals).

The differences between the observed and expected frequencies are:

	Conservative	Liberal Democrat	Labour	Total
Male	−3.645	−11.912	15.556	−0.001
Female	3.645	11.912	−15.556	0.001
Total	0	0	0	0

Each of the row and column totals for these differences should be zero (to an appropriate accuracy). Use this to check your calculations.

The value of the test statistic, X^2 is calculated as follows:

$$X^2 = \frac{(-3.645)^2}{316.645} + \cdots + \frac{15.556^2}{375.444} + \frac{3.645^2}{340.355} + \cdots + \frac{(-15.556)^2}{403.556}$$

$$= 3.34 \text{ (to 2 dp)}$$

To complete the test, the value of X^2 is compared with the upper tail percentage point of the relevant χ^2 distribution.

This test is usually called the chi-squared test for independence.

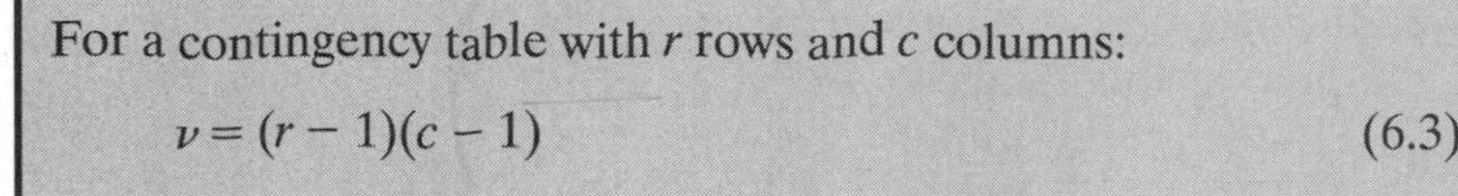

For a contingency table with r rows and c columns:

$$\nu = (r - 1)(c - 1) \quad (6.3)$$

To see the reason for this look again at the expected frequencies:

316.64	135.91	?	828
?	?	?	890
657	282	779	1718

After calculating the $(r - 1)(c - 1)$ (= 2) expected frequencies that are shown, the remainder are not 'free' – their values are fixed by the need for them to sum to the known row and column totals.

Using the 5% significance level with the data, the null hypothesis of independence is accepted, since the observed value (3.34) does not exceed the upper 5% point of a χ_2^2 distribution (5.991).

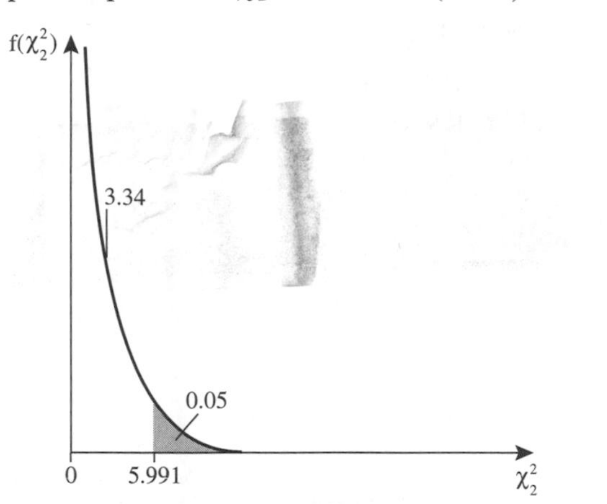

In other words, the voting patterns of the males and females do not differ significantly (at the 5% level).

The conclusion is that 'voting' and 'gender' are independent.

Calculator practice

There are a lot of calculations needed for this test and it is easy to make a mistake. However, some calculators are able to perform these calculations routinely, and if yours is one of these, then you may wish to learn how to use it. You can check whether you are using it correctly by performing the calculations for the previous example.

Typically, you start by entering the data row-by-row (or column-by-column!) into the calculator (the manual may refer to the data as forming a 'matrix'). You will then need to perform a short sequence of key presses with selections between alternatives (see your manual for details).

After the appropriate key sequence, the calculator will typically display the value of X^2, the value of 'df' (= ν) and a 'p-value', which is $P(\chi_\nu^2 > X^2)$. You must then decide the result of the test by comparing the p-value with the given significance level. Thus the null hypothesis of independence would be accepted, at the 5% level, if $p > 0.05$ and rejected if $p \leqslant 0.05$.

Some calculators can also display the table of expected frequencies. You should learn how to do this, since you need to check that all expected frequencies are greater than 5.

If you use a calculator you should check carefully that you have input the data correctly. In examinations a wrong answer with no written working could score no marks.

Example 3

The following data refer to visits to patients in a psychiatric hospital.

Visiting frequency	Length of stay in hospital (years) 2–10	10–20	>20	Total
Regular	43	16	3	62
Sometimes	6	11	10	27
Never	9	18	16	43
Total	58	45	29	132

Verify that there is significant evidence, at the 0.5% level, of a lack of independence (an 'association') between length of stay and the frequency with which a patient is visited.

Here:

H_0: 'Frequency of visiting' and 'length of stay' are independent.
H_1: These characteristics are not independent of one another.

Since there are 3 rows and 3 columns, the relevant chi-squared distribution has $(3 - 1) \times (3 - 1) = 4$ degrees of freedom. If the characteristics were independent, then the expected frequencies would be given by the

$$\frac{\text{Row total} \times \text{Column total}}{\text{Grand total}}$$

formula. Thus the expected frequency for regular visits to recent patients is $\frac{62 \times 58}{132} = 27.24$ (to 2 dp). The complete set of expected frequencies would be as follows:

27.24	21.14	13.62	62.00
11.86	9.20	5.93	26.99
18.89	14.66	9.45	43.00
57.99	45.00	29.00	131.99

Three totals end in 0.99. This is due to round-off errors resulting from giving just two decimal places. The true totals equal those for the observed frequencies. The effect on X^2 is negligible.

Although there is an observed frequency less than 5, there is no need to combine categories, since all the expected frequencies are greater than 5.

The calculations to determine the value of X^2 are shown in the following table:

O_i	E_i	$O_i - E_i$	$\frac{(O_i - E_i)^2}{E_i}$
43	27.24	15.76	9.11
16	21.14	−5.14	1.25
3	13.62	−10.62	8.28
6	11.86	−5.86	2.90
11	9.20	1.80	0.35
10	5.93	4.07	2.79
9	18.89	−9.89	5.18
18	14.66	3.34	0.76
16	9.45	6.55	4.55
Totals: 132	131.99	0	35.17

Your calculator may give a p-value of 4.3×10^{-7}, which is very much less than 0.005, implying that the null hypothesis (of independence) should be rejected in favour of the alternative that there is an association between length of stay and visiting frequency.

Since the value (35.17) of X^2 greatly exceeds the upper 0.5% point of a χ^2_4 distribution (14.860), it is clear that there is a strong association between the classifying variables.

It is often useful to set out the individual cell contributions to X^2 in a table like that of the data, since this helps to show up any pattern in the lack of fit.

Contributions to X^2		
9.11	1.25	8.28
2.90	0.35	2.79
5.18	0.76	4.55

For this data set the major lack of fit arises from the four corner cells – the NW and SE corners have much larger frequencies than would have been expected if there had been no association. Correspondingly, the frequencies in the NE and SW corners are much smaller than would have been expected. These result from the regular visiting of recently admitted patients and the infrequent visits to the very long-stay patients.

Remember that when there are expected frequencies less than or equal to 5 it will be necessary to combine categories. In choosing which categories to combine, you should try to combine neighbours. For example, consider testing the null hypothesis that county of origin and gender are independent using the following data:

County	Males	Females	Total
Cornwall	4	8	12
Devon	6	2	8
Kent	11	10	21
Essex	7	7	14
Suffolk	2	3	5
Total	30	30	60

Since the column totals are equal, the two expected frequencies in each row are equal to half the row total. The expected frequencies for Devon (4 and 4) and for Suffolk (2.5 and 2.5) are less than 5. If the frequencies for these two counties were combined, then all the expected frequencies would exceed 5, as required.

However, this would not be a sensible choice from the geographical viewpoint: Devon and Suffolk are at opposite sides of England. Although it reduces the number of categories by two, rather than by one, the appropriate combinations are Devon with Cornwall and Suffolk with Essex, since these are geographically neighbouring pairs of counties.

Similarly, suppose that one of the categories is age, and that the frequencies in the youngest and oldest age groups are very low. These two categories should not be combined with each other, since they refer to very different ages. Instead the youngest group should be combined with the next youngest group, and the oldest group should be combined with the next oldest group.

S2

Example 4

The political inclinations of a random sample of students from two subject areas are given in the table below, which shows the number of students from each area supporting each political party.

	Arts	Sciences
Conservatives	15	19
Greens	3	7
Labour	22	14
Liberal Democrats	15	20
Others	4	2

Test, at the 2.5% level, the hypothesis that political inclination is independent of subject area.

The row totals are 34, 10, 36, 35 and 6 (total 121).
The column totals are 59 and 62 (total 121)
The resulting table of expected frequencies is (to 2 dp):

	Expected Frequencies	
Conservatives	16.58	17.42
Greens	4.88	5.12
Labour	17.55	18.45
Liberal Democrats	17.07	17.93
Others	2.93	3.07

If you have to calculate the row and column totals, then always check that the totals of these groups of totals are equal to one another (here equal to 121). If they do not agree, then you have made an error.

You can see from the row totals that there are several expected frequencies less than 5. This means that some categories must be combined. The obvious candidates are 'Greens' and 'Others'. Calling the combined group 'Minor parties' gives:

	Arts	Sciences	Total
Conservatives	15	19	34
Labour	22	14	36
Liberal Democrats	15	20	35
Minor parties	7	9	16
Total	59	62	121

S2

The tables of expected frequencies and of (observed − expected) differences are:

Expected frequencies

16.58	17.42
17.55	18.45
17.07	17.93
7.80	8.20

$O - E$

−1.58	1.58
4.45	−4.45
−2.07	2.07
−0.80	0.80

The value of X^2 is given by:

$$X^2 = \frac{(-1.58)^2}{16.58} + \cdots + \frac{0.80^2}{8.20} = 3.14 \text{ (to 2 dp)}$$

After the amalgamation there are 4 rows and 2 columns. Thus there are $(4 - 1) \times (2 - 1) = 3$ degrees of freedom. The upper 2.5% point of a χ_3^2 distribution is 9.348. Since 3.14 is much smaller than 9.348, the null hypothesis that political inclination is independent of subject area is accepted.

Your calculator should give a *p*-value of 0.371 which exceeds 0.025, leading to acceptance of H_0.

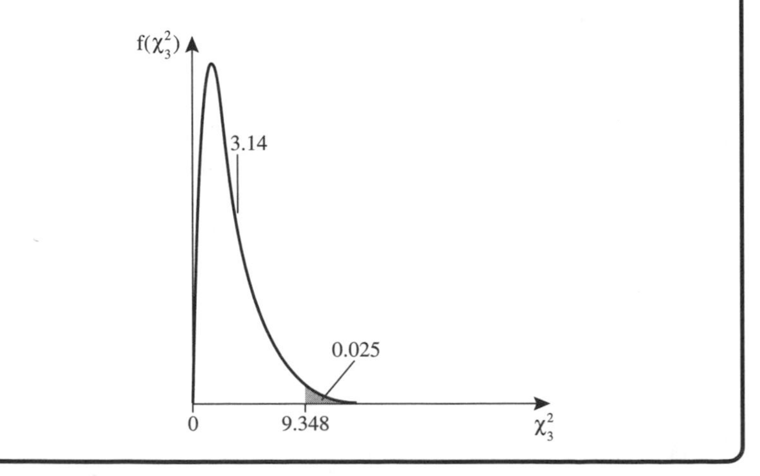

Exercise 6B

1 A survey of the effectiveness of three hospitals in treating a particular illness revealed the following results.

		Complete recovery	Partial recovery	No recovery
Hospital	A	37	23	7
	B	52	44	12
	C	22	30	13

Do the data reveal significant evidence, at the 5% level, of differences in the effectiveness of the hospitals?

2 A survey obtains the following information concerning the party supported by a voter and the greatest academic attainment of the voter.

	Con.	Lab.	Lib.
GCSE	62	111	38
A-level	57	53	25
Degree	24	15	20

Is there significant evidence, at the 1% level, of an association between greatest academic attainment and party supported?

3 A random sample of individuals are classified by age and gender:

Age	Male	Female
16–17	3	2
18–24	15	14
25–60	147	177
61–75	24	30
76–	4	12

Test whether there are significant differences, at the 5% level, between the age distributions of the males and females.

4 The table opposite shows the results of random samples of learner drivers at three driving schools X, Y and Z. The entries, which refer to the success of the drivers at passing the driving test, are percentages of the number of drivers sampled from each school.

	X	Y	Z
% Pass	70	50	40
% Fail	30	50	60

The question of interest is whether the proportion who pass the driving test is independent of the driving school used.

a) Explain why it would be incorrect to calculate the value of the chi-squared test statistic using these percentages.

b) Given that the numbers sampled for X, Y and Z were, respectively, 30, 48 and 20, test at the 5% level of significance, whether there is any evidence of association between the school and the pass-rate. State the null hypothesis.

Give your conclusion, explaining clearly what it means.

5 A university department recorded the A-level grade in a particular subject and the class of degree obtained by 120 students in a given year. The data are summarised in the following table.

		Class of degree		
		I	II	III
A level grade	*A*	10	19	11
	B	4	24	8
	C	2	22	10
	D or *E*	1	5	4

The department wishes to examine the hypothesis that A-level grade and class of degree are independent of one another.

a) Explain why it is necessary to combine the classes '*C*' and '*D* or *E*' into a single class.

S2

b) Having combined these classes, carry out the test, at the 1% level, stating your conclusions clearly.

6 A hospital employs a number of visiting surgeons to undertake particular operations. If complications occur during or after the operation the patient has to be transferred to a larger hospital nearby where the required back-up facilities are available.

A hospital administrator, worried by the effects of this on costs, examines the records of three surgeons. Surgeon A had 6 out of the last 47 patients transferred, surgeon B had 4 out of the last 72 patients transferred, and surgeon C had 14 out of the last 41 transferred.

a) Form the data into a 2×3 contingency table and test, at the 5% significance level, whether the proportion transferred is independent of the surgeon.

b) The administrator decides to offer as many operations as possible to surgeon B.
   i) Give a possible reason.
   ii) Suggest what further information you would need before deciding whether the administrator's decision was based on valid evidence.

7 A market research organisation interviewed a random sample of 125 shoppers in a London supermarket and found that 42 preferred brand *X* washing powder, 63 preferred brand *Y*, and the remainder preferred brand *Z*. A similar survey was carried out in Birmingham. In this survey, of 90 people interviewed, 21 preferred brand *X*, 41 preferred brand *Y* and the remainder preferred brand *Z*.

a) Express the data as a 3×2 table.

b) Test whether these results provide significant evidence, at the 5% level, of different preferences in the two cities.

Yates' correction

In the special case of a 2×2 table, where the approximating χ^2 distribution has just 1 degree of freedom, the approximation is improved by making the small adjustment suggested by Frank Yates in 1926. Denoting the four cells of the table with the suffices 1, 2, 3 and 4, the statistic X_c^2 is used in place of X^2. This revised statistic is defined by:

$$X_c^2 = \sum_{i=1}^{4} \frac{(|O_i - E_i| - 0.5)^2}{E_i} \qquad (6.4)$$

Remember that $|x|$ is always non-negative: $|3| = 3$; $|-2| = 2$.

Here is a simple example:

Values of O_i

10	20	30
30	40	70
40	60	100

Values of E_i

12	18	30
28	42	70
40	60	100

In all 2×2 tables the value of $|O_i - E_i|$ is the same for every cell. This is also true for the value of $|O_i - E_i| - 0.5$.

Thus:

$O_i - E_i$

−2	2
2	−2

$|O_i - E_i| - 0.5$

1.5	1.5
1.5	1.5

Here $|O_i - E_i| - 0.5 = 1.5$.

Since $|O_i - E_i| - 0.5 = 1.5$ for each cell:

$$X_c^2 = 1.5^2 \times \left(\frac{1}{12} + \frac{1}{18} + \frac{1}{28} + \frac{1}{42}\right)$$

An alternative general expression to equation (6.4) is:

$$X_c^2 = (|O_1 - E_1| - 0.5)^2\left(\frac{1}{E_1} + \frac{1}{E_2} + \frac{1}{E_3} + \frac{1}{E_4}\right)$$

This is because, for all i, $|O_i - E_i| = |O_1 - E_1|$

There are several other equivalent expressions for X_c^2. The following alternative avoids calculating the expected frequencies. Denoting the four observed values by a, b, c and d, the marginal totals by m, n, r and s and the grand total by $N(= m + n = r + s)$,

a	b	m
c	d	n
r	s	N

the expression simplifies to:

$$X_c^2 = \frac{N}{mnrs}\left(|ad - bc| - \frac{N}{2}\right)^2 \qquad (6.5)$$

Calculator practice

If you have a calculator that has special functions for the calculation of X^2 for contingency tables, then you should study the instructions with care, to see whether, in the case of a 2×2 table it uses Yates' correction. If this is not the case, then you can use the following transformation:

$$X_c^2 = \left(\frac{D}{O_1 - E_1}\right)^2 X^2$$

where:

$$D = |O_1 - E_1| - 0.5$$

Example 5

The following data comes from a study concerning a possible cure for the common cold. A random sample of 279 French skiers were divided into two groups. Both groups took a pill each day. The pill taken by one group contained the possible cure, whereas the other group took an identical-looking pill that contained only sugar.

Of the 139 skiers taking the possible cure, 17 caught a cold. Of the 140 skiers taking the sugar pill, 31 caught a cold.

Does this provide significant evidence, at the 5% level, that the possible cure has worked?

The hypotheses are:

H_0: The outcome (cold or not) is independent of the treatment (sugar or cure).

H_1: The two characteristics are not independent.

A summary of the observed and expected frequencies is given below:

S2

	Observed Cold	Observed No cold	
Sugar	31	109	140
Cure	17	122	139
	48	231	279

	Expected Cold	Expected No cold	
Sugar	24.086	115.914	140.000
Cure	23.914	115.086	139.000
	48.000	231.000	279.000

Since there are just 2 rows and 2 columns, Yates' correction is needed. We could calculate X_c^2 using:

$$X_c^2 = \frac{(|31 - 24.086| - 0.5)^2}{24.086} + \dots + \frac{(|122 - 115.086| - 0.5)^2}{115.086}$$

$$= (6.914 - 0.5)^2 \times \left(\frac{1}{24.086} + \frac{1}{115.914} + \frac{1}{23.914} + \frac{1}{115.086}\right)$$

$$= 4.14$$

Alternatively, avoiding the possible round-off errors that might result from calculating the expected frequencies, you could use:

$$X_c^2 = \frac{279}{140 \times 139 \times 48 \times 231} \times \left(|(31 \times 122) - (109 \times 17)| - \frac{279}{2}\right)^2$$

$$= 4.14$$

Since 4.14 is a little greater than the upper 5% point (3.841) of a χ_1^2 distribution, H_0 is rejected in favour of H_1. There is evidence (at the 5% level) that the cure has worked!

However remember that, for a significance test at the 5% level, P(Type I error) = 0.05.

A sophisticated calculator with a chi-squared facility should give $X^2 = 4.81$. The corrected value is then obtained by calculating $X_c^2 = \left(\frac{6.414}{6.914}\right)^2 \times 4.81 = 4.14$

Exercise 6C

1 A survey of the traffic passing along a particular road concentrates on the age of the car and the sex of the driver. The results are as follows.

	New car	Old car
Male	117	63
Female	52	48

Using a 5% significance level, test whether there is an association between the sex of the driver and the age of the car.

2 A Danish survey investigated attitudes towards early retirement. Of 317 people in bad health, 276 were in favour of early retirement. Of 258 in moderate health, 232 were in favour. Of 86 in excellent health, 73 were in favour. All those not in favour were against the idea.

a) Express the data as a 3×2 table.

b) Test the hypothesis that attitudes towards early retirement are independent of health.

c) Repeat the test ignoring the data on those who were in moderate health. State your conclusions.

3 As part of a research study into pattern recognition, subjects were asked to examine a picture and see if they could distinguish a word. The picture contained the word 'scitamehtam' camouflaged by an elaborate pattern. Of 23 librarians who took part, 11 succeeded in recognising the word, whilst of 19 statisticians, 13 succeeded. Form the data into a 2×2 contingency table and test at the 5% significance level, whether an equal proportion of librarians and statisticians can distinguish the word. You may assume that those taking part were random samples from their respective populations.

S2

Summary

You should now be able to ...	Check out
1 Use tables of the χ^2 distribution.	**1** a) If $X^2 \sim \chi^2_9$, find c such that $P(X^2 > c) = 0.05$ b) If $X^2 \sim \chi^2_{4,}$ find c such that $P(X^2 > c) = 0.01$ c) If $X^2 \sim \chi^2_1$, find c such that $P(X^2 > c) = 0.10$.
2 Analyse a contingency table for independence.	**2** Test, at the 5% significance level, the hypothesis that the table displays frequencies consistent with independence. 23 30 47 39 33 53 22 27 26

3 Apply Yates' correction.

3 Determine whether the following table is consistent with the hypothesis of independence using a 1% significance level.

39	21
36	4

S2

Revision exercise 6

1 For her 17th birthday present, Susan wishes to have a course of driving lessons. In an attempt to select the best driving school in the area, her parents compare recent test results of two schools A and B.

These test results are tabulated opposite.

	School A	School B
Pass	120	100
Fail	24	36

Stating the null hypothesis, use a χ^2 test at the 5% level of significance to determine whether there is an association between test results and driving school.

(*AQA, 2002*)

2 The prisoners in a high security jail are placed into three categories: Short stay, Medium stay and Long stay. The frequency with which the prisoners receive mail is monitored and the results recorded in the table below.

	Length of stay in prison			
	Short	Medium	Long	
Regular mail	40	24	8	72
Occasional mail	12	18	15	45
No mail	13	27	22	62
	65	69	45	179

Stating your null and alternative hypotheses, investigate, at the 1% level of significance, the claim that there is no association between length of stay in prison and the frequency with which prisoners receive mail.

(*AQA, 2002*)

3 The results of a recent police survey of traffic travelling on motorways produced information about the genders of drivers and the speeds, S miles per hour, of their vehicles, as tabulated below.

	Speed of vehicle		
	$S \leqslant 70$	$70 < S \leqslant 90$	$S > 90$
Male	17	40	70
Female	30	25	18

Stating null and alternative hypotheses, investigate, at the 1% level of significance, the claim that there is no association between the gender of the driver and the speed of the vehicle.

(*AQA, 2003*)

4 Baljeet's parents want to choose a school for her to attend in order to pursue her sixth form studies. They have a choice of sending her to one of two schools, X or Y. To help them make an informed choice, they decide to look at the numbers of A, B, C and D grades achieved last year by each school. These are tabulated below.

	Examination Grades				
	A	B	C	D	Total
School X	52	34	16	18	120
School Y	114	58	62	46	280
Total	166	92	78	64	400

Stating the null hypothesis, carry out a χ^2 test at the 5% level of significance to determine whether there is an association between the schools and the numbers of A, B, C and D grades achieved. *(AQA, 2004)*

5 Students following a media studies course at a particular university must study a foreign language. They are given a choice of Japanese, Russian or Swedish. The following table summarises the choices by gender of a random sample of students entering the course.

	Japanese	Russian	Swedish
Female	16	12	22
Male	12	14	54

a) Use the χ^2 distribution and the 5% significance level to investigate whether choice of language is associated with gender.

b) i) Interpret your result in part a) as it relates to the choice of Swedish.

ii) Make a further statement about the choice of Swedish. *(AQA, 2002)*

6 A railway company wishes to monitor passenger opinion. During a pilot study a randomly selected sample of rail passengers was surveyed. As part of the survey they were asked whether they carried mobile phones for use on their journey and whether they were irritated by other passengers using mobile phones on the journey.

The results are summarised in the following table.

	Irritated	Not irritated
Mobile phone carried	12	14
Mobile phone not carried	26	6

a) i) Use a χ^2 test, at the 5% significance level, to analyse the contingency table above.

ii) Interpret your conclusion in the context of the question.

b) As a result of the pilot study the questionnaire was modified and a further random sample of rail passengers was asked to complete it. The answers to the questions relating to mobile phones are summarised in the following table.

	Not irritated	Slightly irritated	Irritated	Extremely irritated
Mobile phone carried	6	46	18	3
Mobile phone not carried	3	7	14	17

The expected values for the analysis of this contingency table are given below.

	Not irritated	Slightly irritated	Irritated	Extremely irritated
Mobile phone carried	5.76	33.94	20.49	12.81
Mobile phone not carried	3.24	19.06	11.51	7.19

Jason is asked to analyse the data and suggests the 'Not irritated' and 'Extremely irritated' columns be combined together before carrying out the analysis.

Eric suggests that it would be better to combine the 'Not irritated' column with the 'Slightly irritated' column and the 'Extremely irritated' with the 'Irritated' column.

Anthony suggests that the 'Not irritated' column should be combined with the 'Slightly irritated' column and the resulting 2×3 table analysed.

i) Give one criticism of Jason's suggestion.

ii) Explain why Anthony's suggestion is preferable to Eric's.

iii) Carry out Anthony's suggestion, using the 1% significance level, and interpret your conclusion in context.

(AQA, 2002)

7 Rhamesh runs a café which opens at noon and admits no further customers after 9 pm. He does not sell alcoholic drinks, but customers can bring their own beer or wine with them, to drink with their meal, if they wish.

To help him decide whether it would be worthwhile selling his own alcoholic drinks, Rhamesh asks Anna, a student, to record how many customers bring their own alcoholic drinks. Anna suspects this may be related to the time of day, and following her observations she produces the table below.

The table shows the number of customers, with and without alcoholic drinks, who entered the café at different times on a particular day.

	Time of Day		
	Noon to 3 pm	3 pm to 6 pm	6 pm to 9 pm
Alcoholic drink	12	6	54
No alcoholic drink	62	35	67

a) Use the 1% significance level to investigate whether taking alcohol with a meal is independent of time of day.

b) i) Interpret your result in part a) as it relates to customers entering the café between 6 pm and 9 pm.

ii) Make a further statement about customers entering the café between 6 pm and 9 pm.

c) How might your conclusions be affected, if at all, by the fact that the table shows the customers who entered the café on only one particular day? (*AQA, 2003*)

8 A well-known picture of the Beatles shows them on a pedestrian crossing outside the Abbey Road recording studios. Substantial numbers of Beatles fans visit Abbey Road and use the pedestrian crossing.

It was claimed that these fans were causing delays to rush hour traffic. As part of an investigation, people using the crossing were asked whether they were using it in the course of their normal daily lives or because of the Beatles photograph. The answers and the times of day are summarised in the contingency table.

	Time of Day	
	Rush hour	Out of rush hour
Normal daily lives	34	45
Because of Beatles	4	28

a) Using the 5% significance level, examine whether the reason for using the crossing is associated with the time of day.

b) Interpret your result in the context of this question. (*AQA, 2003*)

9 Chandra, an estate agent, sells properties in a large conurbation. She classifies the areas of the conurbation as:

City Centre;
Inner City – close to but outside the City Centre;
Suburban – outside the Inner City.

The table shows the number of properties she has for sale in each of these areas together with information on the Advertised Price.

		Area		
		City Centre	Inner City	Suburban
Advertised Price	<£120 000	2	58	20
	⩾£120 000	34	12	44

a) Use a χ^2 distribution and the 1% significance level to analyse this contingency table.

b) Interpret your result in the context of the question.

c) Chandra's colleague Michael suggests that a better way of analysing the relationship between price and area would be to calculate the product moment correlation coefficient between the advertised price of a property and its distance from the centre of the city.

Comment on this suggestion.

i) in general terms;

ii) as it relates to your conclusions in part b). (*AQA, 2004*)

S2

MS2A Practice Paper (with coursework)

75 minutes *60 marks* *You may use a graphics calculator.*

*Answer **all** questions.*

S2

1 A machine is used to produce thin metal discs which have thicknesses, in millimetres, that are normally distributed with mean μ and variance σ^2. A quality control inspector measures the thickness, x millimetres, of each of a random sample of 10 discs with the following results.

1.22 1.34 1.46 1.31 1.28 1.17 1.38 1.37 1.41 1.26

(a) Calculate unbiased estimates of μ and σ^2. *(2 marks)*

(b) Hence construct a 99% confidence interval for μ, giving the limits to two decimal places. *(5 marks)*

2 The discrete random variable R has the following probability distribution.

$$P(R = r) = \begin{cases} \dfrac{r}{15} & r = 1, 2, 3, 4, 5 \\ 0 & \text{otherwise} \end{cases}$$

(a) Calculate values for the mean and variance of R. *(5 marks)*

(b) A circle is drawn with radius $(3R - 2)$ centimetres.

Determine, in terms of π, values for the mean and variance of the circumference of the circle. *(3 marks)*

3 A bakery produces loaves of bread which, according to the specification, should have a mean weight of 415 grams. A random sample of 100 loaves, taken from a large batch, had a mean weight of 413.9 grams and a standard deviation of 7.4 grams.

(a) Investigate, at the 5% level of significance, the suspicion that the mean weight of loaves in this batch is below the specification. *(6 marks)*

(b) Explain, in the context of this question, the meaning of a Type II error. *(2 marks)*

4 The views of a random sample of 250 private shareholders of a company were sought about a proposed takeover of the company. The results are shown in the contingency table.

	View on proposal	
Share holding	**In favour**	**Against**
Less than 100	7	3
100 to 999	83	17
1000 to 9999	57	33
At least 10 000	28	22

(a) Calculate values for the 8 expected frequencies. *(4 marks)*

(b) Explain why combining some cells is necessary before a χ^2 test can be undertaken. *(2 marks)*

(c) Investigate, at the 5% level of significance, a claim that there is an association between share holding and attitude to the proposal. *(6 marks)*

(d) Interpret, in context, your conclusion to part (c). *(2 marks)*

5 A process used for manufacturing sheets of glass produces small bubbles, called 'seeds', scattered at random in the glass at an average rate of 0.2 per square metre of glass.

(a) Two panes of glass, each 2 m $\times$ 3 m, are inspected for seeds. Find the probability that:
(i) the first pane inspected contains fewer than 2 seeds; *(2 marks)*
(ii) the total number of seeds in the two panes is exactly 4. *(3 marks)*

(b) A pane of glass is to be cut, with an area of k square metres, so that the probability of the pane containing no seeds is at least 0.8.

Calculate, to three decimal places, the largest possible value of k. *(4 marks)*

6 A tennis player hits a ball against a wall, aiming at a fixed horizontal line on the wall. The vertical distance from the horizontal line to the point where the ball strikes the wall is recorded as positive for points above the line and negative for points below the line.

It is assumed that the distribution of this vertical distance, X metres, may be modelled by the probability density function:

$$f(x) = \begin{cases} 1.5(1 - 4x^2) & -0.5 \leqslant x \leqslant 0.5 \\ 0 & \text{otherwise} \end{cases}$$

(a) Sketch the graph of f. *(3 marks)*

(b) State the probability that the ball strikes the wall precisely 0.25 metres above the line. *(1 mark)*

(c) Determine the probability that the ball strikes the wall more than 0.25 metres from the line. *(4 marks)*

(d) (i) State why $E(X) = 0$. *(1 mark)*
(ii) Calculate $\text{Var}(X)$. *(3 marks)*

(e) Give a reason why the above probability model may **not** be appropriate. *(1 mark)*

(f) Suggest one likely effect of repeated practice on the above probability model. *(1 mark)*

S2

MS2B Practice Paper (without coursework)

90 minutes *75 marks* *You may use a graphics calculator.*

*Answer **all** questions.*

1 A machine is used to produce thin metal discs which have thicknesses, in millimetres, that are normally distributed with mean μ and variance σ^2. A quality control inspector measures the thickness, x millimetres, of each of a random sample of 10 discs with the following results.

1.22 1.34 1.46 1.31 1.28 1.17 1.38 1.37 1.41 1.26

(a) Calculate unbiased estimates of μ and σ^2. *(2 marks)*

(b) Hence construct a 99% confidence interval for μ, giving the limits to two decimal places. *(5 marks)*

2 The discrete random variable R has the following probability distribution.

$$P(R = r) = \begin{cases} \frac{r}{15} & r = 1, 2, 3, 4, 5 \\ 0 & \text{otherwise} \end{cases}$$

(a) Calculate values for the mean and variance of R. *(5 marks)*

(b) A circle is drawn with radius $(3R - 2)$ centimetres.

Determine, in terms of π, values for the mean and variance of the circumference of the circle. *(3 marks)*

3 The random variable X has a rectangular distribution on the interval $-2c$ to $+4c$, where c is a positive constant.

(a) Find, in terms of c, the mean and variance of X. *(2 marks)*

(b) Hence, given that $E(X^2) = 16$, show that $c = 2$. *(2 marks)*

(c) Given that $P(-b < X < 5b) = 0.25$, find the value of b. *(2 marks)*

S2

4 The views of a random sample of 250 private shareholders of a company were sought about a proposed takeover of the company. The results are shown in the following contingency table.

	View on proposal	
Share holding	**In favour**	**Against**
Less than 100	7	3
100 to 999	83	17
1000 to 9999	57	33
At least 10 000	28	22

(a) Calculate values for the 8 expected frequencies. *(4 marks)*

(b) Explain why combining some cells is necessary before a χ^2 test can be undertaken. *(2 marks)*

(c) Investigate, at the 5% level of significance, a claim that there is an association between share holding and attitude to the proposal. *(6 marks)*

(d) Interpret, in context, your conclusion to part (c). *(2 marks)*

5 Alan and Bruno go fishing every Sunday. The number of fish caught by each man may be modelled by a Poisson distribution, with a mean of 3.4 for Alan and a mean of 2.6 for Bruno. On any Sunday, the number of fish caught by Alan is independent of the number of fish caught by Bruno.

(a) For a Sunday chosen at random, determine the probability that:

(i) **in total**, the two men catch more than 6 fish; *(3 marks)*

(ii) the men **each** catch exactly 3 fish. *(3 marks)*

(b) For a randomly chosen Sunday, determine the probability that at least one of the two men fails to catch any fish. *(3 marks)*

(c) Given that exactly 6 fish are caught on a randomly chosen Sunday, determine the probability that they are all caught by Alan. *(4 marks)*

6 A tennis player hits a ball against a wall, aiming at a fixed horizontal line on the wall. The vertical distance from the horizontal line to the point where the ball strikes the wall is recorded as positive for points above the line and negative for points below the line.

It is assumed that the distribution of this vertical distance, X metres, may be modelled by the probability density function:

$$f(x) = \begin{cases} 1.5(1 - 4x^2) & -0.5 \leq x \leq 0.5 \\ 0 & \text{otherwise} \end{cases}$$

(a) Sketch the graph of f. *(3 marks)*

(b) State the probability that the ball strikes the wall precisely 0.25 metres above the line. *(1 mark)*

(c) Determine the probability that the ball strikes the wall more than 0.25 metres from the line. *(4 marks)*

(d) (i) State why $E(X) = 0$. *(1 mark)*
(ii) Calculate $Var(X)$. *(3 marks)*

(e) Give a reason why the above probability model may **not** be appropriate. *(1 mark)*

(f) Suggest one likely effect of repeated practice on the above probability model. *(1 mark)*

7 When used to record the viscosity of a liquid, an instrument provides measurements that are normally distributed about the true value with a standard deviation of 0.60 pascal seconds.

The viscosity of oil in a tank is to be checked by calculating the mean, $\overline{X}$ pascal seconds, of a random sample of 9 independent measurements to be taken using the instrument.

(a) State hypotheses to investigate the claim that the viscosity of oil in the tank exceeds the specification of 34.20 pascal seconds. *(1 mark)*

(b) The significance level of the hypothesis test is to be 5%.

State what is meant by this statement. *(1 mark)*

(c) State why $\overline{X}$ has a normal distribution. *(1 mark)*

(d) Determine the critical region for the hypothesis test in the form $\overline{X} > k$, giving the value of k to two decimal places. *(4 marks)*

(e) Explain, in the context of this question, what is meant by this critical region. *(2 marks)*

(f) Given that the value of the sample mean was 34.45 pascal seconds, state, with a reason, the conclusion you would reach. *(2 marks)*

(g) Explain, in the context of this question, what is meant by a Type II error. *(2 marks)*

7 Coursework guidance

This chapter is for students taking the S2A unit.

The S2B unit does not contain coursework.

If you are unsure which unit you are taking, you should ask your teacher.

In this chapter you will find:

- A clear description of how to tackle your statistics coursework
- A strand by strand breakdown of the marking grid specifically geared to the A2 tasks
- Useful tips and hints from experienced moderators
- Answers to some frequently asked questions
- A final checklist

7.1 Introduction

The S2 coursework task will probably be the second piece of statistics coursework that you attempt in your mathematics course. It should build upon the experiences and skills you developed when tackling the AS-level tasks in the S1A unit. If you have not yet attempted any statistics coursework, do not worry as the guidance in this chapter should enable you to complete successfully the task you choose.

The emphasis of the coursework is on modelling a 'real-life' situation using appropriate statistical distributions. You will need to design a given task, discuss your sampling and data collection methods, perform some relevant calculations and interpret your results.

The task will be assessed at A2 standard so there will be increased expectations of the piece of work you produce. For the highest marks there will need to be a maturity of expression, clarity in your use of statistical terminology and notation, as well as clear and appropriate discussion of your results in the context of the task. Use the skills you have developed in the unit to produce an interesting and relevant write-up which is clear to follow and 'a good read'!

This chapter is to help you with the coursework process right from the starting point to handing in the completed piece of work. There are useful hints and tips from experienced moderators who work for the Examination Board. There is a clear and full description of the marking grid which will be used to assess your piece of work adapted for the type of tasks set at S2A-level.

For your S2A coursework you will need to offer one task. This is worth 25% of the marks available for the unit ($4\frac{1}{6}$% of the total A-level award).

7.2 Choosing a task

The Examination Board will provide a list of tasks which are appropriate for your S2A coursework. Your teacher may decide to offer one task or provide you with a number of choices. It is important to choose a task that you feel comfortable with, and one which gives you the scope to use your statistical skills fully.

Listen carefully to the advice of your teacher. Do not start a task which has not been approved by your teacher.

Just as in your S1A task your teacher may allocate time in class to discuss various ideas and approaches that you might take.

- Write down your ideas
- Discuss these with others in your class
- Adapt and modify your ideas and reject as necessary

Most importantly, learn from any mistakes you made in your approaches to the S1A task. Your teacher should be able to help you with this.

7.3 The coursework process

Your coursework can be broken down into a number of stages. In the A2 task some of these stages may be more important to your task than others, but **all** of them need to be addressed.

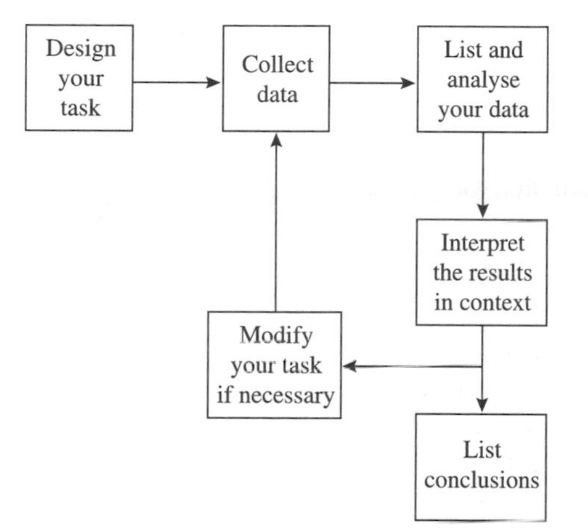

The main focus of the coursework for S2A is to perform statistical tests on the data that you collect.

These could be a small sample t-test for μ, a larger sample z-test for μ or a χ^2 test for independence in a contingency table.

7.4 Assessment Criteria

Statistics coursework is marked under four strands:

- Design
- Data collection and statistical analysis
- Interpretation and validation
- Communication

There are 80 marks in total. The marking grid on the next page shows how the marks are broken down. Although the marking criteria are identical for all statistics tasks, the criteria need to be interpreted in the context of each task.

Marking grid for Statistics S2

Strand	0–8 marks	9–15 marks	16–20 marks	Mark awarded
1. **Design**	Problem defined and understood. Aims and objectives discussed. Some discussion of how the sample was obtained.	The approach to the task is coherent. The strategies to be employed are appropriate. Clear explanation of how sample was obtained. Some discussion of the statistical theories or distributions used.	A well-balanced and coherent approach. Clear discussion and justification of the statistical theories or distributions used in relation to the task.	/20
2. **Data Collection and Statistical Analysis**	Adequate data collected. Raw data clearly set out. Some relevant calculations are correct.	A range of relevant statistical calculations are used. Most calculations are correct, and quoted to an appropriate degree of accuracy.	A full range of calculations are used. The calculations are correct and appropriate to the task.	/20
3. **Interpretation and Validation**	A reasoned attempt is made to interpret the results. Some discussion of how realistic the results are. Some discussion of possible modifications.	Results are interpreted. Attempt to relate the task to the original problem. Clear discussion of possible modifications/ improvements which could have been made.	Results are fully interpreted within a statistical context. Outcomes are clearly related to the original task. Clear discussion of the effects of the sampling and data collection methods used.	/20
4. **Communication**	The report is presented clearly and organised with some explanation. Diagrams are effective and appropriate. Conclusions are stated.	The report is clear and well organised. Other areas of work which could have been investigated are discussed. The report is consistent with a piece of work of 8–10 hours.	Appropriate language and notations used throughout. The report is clear and concise and of sufficient depth and difficulty.	/20
			Total Mark	**/80**

This section will discuss in detail the four strands that you will be assessed on. Reference will be made to the marking grid to help you understand exactly what is expected of you.

Strand 1: Design

✦ **Candidates must state clear aims and identify their strategies and objectives. (4 marks)**

Question: What exactly are you going to do?
It must be clear what the purpose of the piece of work is. It is a good idea to write a short introduction describing the task in your own words along with some initial thoughts about the potential outcomes you might expect. Remember to mention the statistical techniques you intend to use to complete the task.

You will not be given 4 marks for simply copying out the original task as given.

If performing a hypothesis test for the mean mass of a type of chocolate bar, you could choose bars from two different manufacturers to test if the contents are as stated on the wrappers.

If testing for the independence of two variables such as voting intention and newspaper preference, how are you going to split the categories before collecting the data?

✦ **They must clearly define the population being used. (2 marks)**

For a full 2 marks you will need to state clearly the population from which you will be sampling. You will need to explain your choice of population to justify its appropriateness. For the S2A task this may need careful thought.

Is it really sensible to choose a population of 'all chocolate bars in the UK'?

✦ **They must explain how their sample was obtained and give details of any questionnaires/experiments used. (6 marks)**

You will need to choose some appropriate data to meet your aims. The intention is that you will collect your own data to analyse, and for S2A tasks this will give you far more scope for discussion in your write-up.

In some cases it may be necessary to collect data in small groups. If this happens, then the final write-up must be your own work and you will sign a declaration to state that it is.

You must note in your work if the data was collected as a group, and you should state who was in your group.

Question: How are you going to collect your data?
Whatever task you do, it is necessary for you to take a sample from a population. It is therefore important to discuss your sampling method in some detail. The following points should be considered:

- ✦ What sample size will you use and why?
- ✦ Are you trying to choose a random sample where each member of the population has an equal chance of being selected? If so, how are you going to obtain it?

If doing a t-test for μ from a normal population with an unknown variance you can use a relatively small sample size. If you are looking to do a z-test for μ using a normal approximation then you need a larger sample size.

Picking the first 10 chocolate bars in a shop is collecting a sample but is it statistically appropriate?

Practicalities may govern how you collect your sample. If this is so, then discuss it in your write-up explaining why you may have used a non-random method.

If you are collecting data to put into contingency tables then you will almost certainly need to collect the information from either a questionnaire, a survey or by performing a suitable experiment.

How will you measure the masses of the chocolate bars? What happens if a particular bar is broken?

How will you collect the voting intentions of a sample of adults? What happens if a person gives a response which is not in one of your categories?

Remember to consider:

- Who or what you use in your sample
- Problems which could occur when collecting the sample and how you would deal with them
- How you will ensure randomness to minimize bias in your sample selection.

Despite using a random method of selecting your sample you can still end up with an unusual sample. All you are ensuring is that your method of selection is random!

All of these issues need to be discussed in your report as there are 6 marks specifically allocated to your sampling method.

- **There should be a clear discussion of which statistical theories or distributions are being used and why. (6 marks)**

Throughout your coursework you will be introducing various statistical techniques to analyse your data. You need to discuss statistical theories, but only when they arise and in context. Be selective and express the relevant theory as you understand it using your own words.

If you are performing a t-test for μ you may need to discuss:

- The null and alternative hypotheses
- Whether a one or two tailed test is appropriate
- Why the t-test is appropriate for your data
- The significance level
- The number of degrees of freedom
- The test statistic
- How the critical values are obtained from the t-tables

If you are carrying out a contingency table test of independence you may need to discuss:

- ✦ The null and alternative hypotheses
- ✦ How the expected frequencies are calculated
- ✦ What happens if an expected frequency is ≤ 5
- ✦ The test statistic
- ✦ The number of degrees of freedom
- ✦ The significance level
- ✦ The use of Yates' correction if required
- ✦ How the critical values are obtained from the χ^2 tables

Do not copy out vast chunks of text from a book just because you feel it might be relevant to your analysis.

✦ **Assumptions made should be fully discussed** **(2 marks)**

Question: In planning to collect your data, what assumptions or rules are you making?

- ✦ Do you need to make any assumptions about the data for a t-distribution to be valid?
- ✦ Do you need to make any rules about the data that you collect or the experiments that you perform?

State clearly what you are assuming and ensure that your assumptions are valid. If you apply your rules consistently you will satisfy the criteria.

S2

Strand 2: Data Collection and Statistical Analysis

This strand assesses how successful you are in the collection of your data and the quality of the analysis that you have performed on it.

✦ **Raw data must be recorded and then organized as appropriate.** **(2 marks)**

There is an expectation that you will list your data in its original raw form. In some cases it may be sensible to group it into frequency distributions or tables. You can put your data in the body of the work or insert it as an appendix.

✦ **An adequate amount of data should be collected.** **(2 marks)**

If performing a χ^2 test the sample size should be large enough to avoid having many small expected frequencies. For a z-test with σ^2 unknown $n \geq 30$ is advisable, but for a t-test a small sample can be used.

It is worth mentioning that you know these facts to show that you have thought about the issue of sample size.

Remember you should not have a sample which is the whole population.

✦ **The calculations are correct and appropriate to the statistical model and content of the unit.** **(10 marks)**

If you are carrying out a hypothesis test based on a small sample size you may need to calculate:

- ✦ The sample mean
- ✦ The unbiased estimate of the population variance
- ✦ The test statistic
- ✦ A critical value from the appropriate t distribution
- ✦ A confidence interval for μ

Beware of using every statistical technique you have heard of, whether relevant or not.

Here is an example:

You may have collected some data from which you wish to perform a hypothesis test on μ. Suppose that μ is thought to be 20, but after recent adjustments it has been suggested that the value of μ is now greater than 20.

A random sample of 10 observations of the variable X gave the following results:

20.7 20.4 20.6 21.1 19.7 19.9 21.2 21.9 21.4 20.6

H_0: $\mu = 20$
H_1: $\mu > 20$ (one tailed test)

Under H_0 (and assuming that the sample is drawn from a normal population with unknown variance):

$$\frac{\bar{x} - 20}{\sqrt{\frac{s^2}{10}}} \sim t_9$$

From the sample:

$$\bar{x} = \frac{\sum x}{n} = \frac{207.5}{10} = 20.75 \qquad s^2 = \frac{1}{n-1}\left[\sum x^2 - \frac{(\sum x)^2}{n}\right] = \frac{1}{9}\left[4309.69 - \frac{(207.5)^2}{10}\right] = 0.451\dot{6}$$

Calculate the test statistic

$$t = \frac{20.75 - 20}{\sqrt{\frac{0.451\dot{6}}{10}}} = 3.53 \text{ (3 sf)}$$

This value can then be compared to critical values for a one tailed test at various percentage levels for a t_9–distribution. A suitable conclusion can then be made in context.

Note that the critical value at the 5% level for a t_9–distribution is 1.833. Since $3.53 > 1.833$ you would reject H_0. In this case it would be a good idea to proceed by calculating a confidence interval for the value of μ based upon the sample.

If you are focussing on a contingency table test of independence you may need to:

- ✦ Calculate expected frequencies
- ✦ Merge cells if any expected frequencies are $\leqslant 5$
- ✦ Use Yates' correction if a 2×2 table is obtained
- ✦ Calculate the value of X^2
- ✦ Calculate the number of degrees of freedom

S2

Here is an example:

You may have collected some data to test whether the grade achieved in GCSE Physics is independent of the grade achieved in GCSE Mathematics. 200 students were chosen at random and the results categorised as follows:

		Mathematics grade			
		A or A*	B	C	D or worse
Physics grade	A or A*	28	18	18	1
	B	15	25	20	3
	C	6	17	21	1
	D or worse	2	3	2	20

H_0: Mathematics and Physics grades at GCSE are independent
H_1: Mathematics and Physics grades at GCSE are not independent

Calculate row and column totals:

28	18	18	1	65
15	25	20	3	63
6	17	21	1	45
2	3	2	20	27
51	63	61	25	200

Now calculate the expected frequencies:

		Mathematics grade				
		A or A*	B	C	D or worse	
Physics grade	A or A*	$\frac{51 \times 65}{200} = 16.575$	$\frac{63 \times 65}{200} = 20.475$	$\frac{61 \times 65}{200} = 19.825$	$\frac{25 \times 65}{200} = 8.125$	65
	B	$\frac{51 \times 63}{200} = 16.065$	$\frac{63 \times 63}{200} = 19.845$	$\frac{61 \times 63}{200} = 19.215$	$\frac{25 \times 63}{200} = 7.875$	63
	C	$\frac{51 \times 45}{200} = 11.475$	$\frac{63 \times 45}{200} = 14.175$	$\frac{61 \times 45}{200} = 13.725$	$\frac{25 \times 45}{200} = 5.625$	45
	D or worse	$\frac{51 \times 27}{200} = 6.885$	$\frac{63 \times 27}{200} = 8.505$	$\frac{61 \times 27}{200} = 8.235$	$\frac{25 \times 27}{200} = 3.375$	27
		51	63	61	25	200

As $3.375 < 5$ you will need to merge a row or column in the table.

Suppose you collapse columns 3 and 4 to form a 'C or worse' column. This would give:

observed

	A or A*	B	C or worse
A or A*	28	18	19
B	15	25	23
C	6	17	22
D or worse	2	3	22

expected

	A or A*	B	C or worse
A or A*	16.575	20.475	27.95
B	16.065	19.845	27.09
C	11.475	14.175	19.35
D or worse	6.885	8.505	11.61

Calculate the value of X^2.

$$X^2 = \sum \frac{(O_i - E_i)^2}{E_i}$$

$$= \frac{(28 - 16.575)^2}{16.575} + \frac{(18 - 20.475)^2}{20.475} + \ldots + \frac{(22 - 11.61)^2}{11.61}$$

$$= 32.9 \text{ (3 sf)}$$

Number of degrees of freedom $= (r - 1)(c - 1)$

$= (4 - 1)(3 - 1)$

$= 6$

The critical value for a χ_6^2 distribution at the 1% level (say) is 16.812.

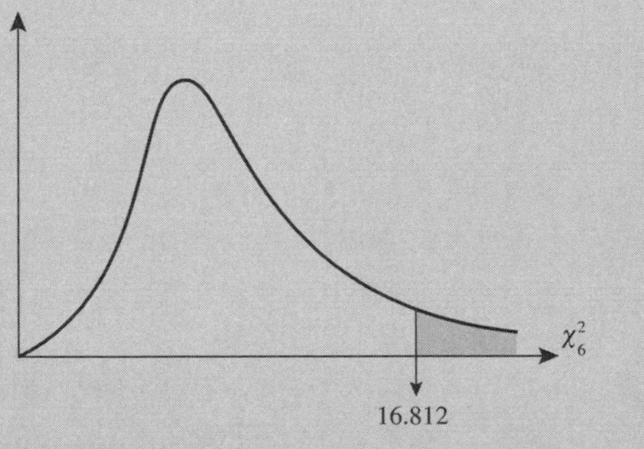

As $32.9 > 16.812$, you reject H_0.

The evidence suggests that Mathematics and Physics grades at GCSE are not independent.

✦ **A sufficient range of relevant calculations and analysis should be used.** **(4 marks)**

Your analysis should contain enough calculations to ensure that your original aims can be met. Aim to do at least two hypothesis tests or at least two contingency tables to help with the interpretation.

✦ **Appropriate degrees of accuracy will have been used in the answers given.** **(2 marks)**

An important consideration is the degree of accuracy you will use in your calculations. For example, if you have collected data to 1 dp, is it appropriate to quote your answers to 2 dp?

Whatever you decide to do, be consistent. If you are consistent, but use inappropriate degrees of accuracy, you will gain some credit.

You might decide to use a greater degree of accuracy during your calculations and then round to a sensible accuracy at the end.

Strand 3: Interpretation and Validation.

In this strand you are trying to state what your results show. Do they support your initial ideas?

✦ **Candidates must look at how realistic their final results are, and, if appropriate, give reasons why the answers are not sensible.** **(6 marks)**

You need to decide how realistic your final results are.

If you have performed hypothesis tests to test claims made by a manufacturer, what do your results show? If you accepted their claims, is this what you would have expected? Manufacturers often have Quality Assurance Departments which regularly test their products to check that stated claims are being satisfied, otherwise there could be legal implications.

If there is no obvious method of validation then you should discuss your results to satisfy yourself that they are sensible, and that there are no errors in your calculations.

If you have tested the independence of two variables, such as performance at GCSE and A2/AS level, what does dependence indicate? Is this a spurious link due to some other factor, or do you think it is a tangible link worthy of further research?

✦ **They should give a clear statistical interpretation of their results, along with suitable conclusions.** **(6 marks)**

For this criterion, you are expected to give a statistical interpretation of your results in context.

If you have accepted a null hypothesis in a hypothesis test, explain clearly the meaning of this. You need to link your result to the significance level of the test and recognise that whatever result you obtain, there is a chance that you might have come to a false conclusion.

This should lead to a discussion of Type I and Type II errors linked specifically to your task.

> Remember:
>
> - Type I error: You have rejected H_0 when in fact it is true
> - Type II error: You have accepted H_0 when in fact it is false.

You can quote the probability of a Type I error, which is just the significance level of the test.

You could also calculate the probability of a Type II error for an appropriate value of μ.

- **They must relate the outcomes to the original task and discuss the effect of their sampling and data collection methods on their results. (4 marks)**

Question: Do your results support your original hypotheses? Use specific evidence from your work to support your conclusions.

You should also look back at your original sampling method:

- Was it random?
- Did you introduce bias into the sample (however unintentionally)?
- How could you have adapted your sampling method to remove any bias?
- If you carried out an experiment or collected a questionnaire, was there anything you could have done which could have improved the data collected?

> Simply stating that you could have used a larger sample size is worthy of little credit.

- **They must look at what modifications/improvements could have been made with the benefit of hindsight, (but they should not have to make them unless their analysis is so over-simplified that this becomes necessary). (4 marks)**

This is the opportunity to say if you felt that your design was not everything you hoped it would be!

Question: Could you have improved the quality of the data obtained by changing your original approach?

You only need to suggest alternative approaches, not implement them, unless you have done such a limited analysis that your work lacks depth.

> Your teacher should be able to advise whether you have done sufficient analysis of the task.

It is fine to be self-critical in this section.

Strand 4: Communication

This strand measures how well you have communicated your work in the final write-up.

✦ **Candidates must express themselves clearly and concisely using appropriate mathematical language notation and language. Graphs and diagrams should be clearly and accurately labelled.** **(4 marks)**

This is an overall judgement of how clearly you have expressed your ideas. It is important to be concise, and to be efficient in your presentation.

You do not need to include page after page of repetitive calculations, but you must include at least one worked example of all types of calculations that you have used in your coursework. It is important that these calculations can be checked from the original raw data.

Use the appropriate mathematical terms and notation for the specification. For example you may need to use:

✦ $\bar{x}$ (sample mean)
✦ s^2 (unbiased estimate of the population variance)
✦ μ (population mean)
✦ σ^2 (population variance)
✦ χ^2 (chi-squared distribution)

Include all appropriate diagrams in your write-up.

You may need to draw an appropriate χ^2 distribution, or a *t*-distribution or a normal distribution.

✦ **Candidates should include other areas of work that could have been investigated further.** **(2 marks)**

You are expected to suggest other work that could follow from your coursework.

Question: Where could you take your research further?

If you have examined a link between GCSE score and AS/A2 performance, you could suggest linking IQ to examination performance.

✦ **The final conclusions should be set out logically.** **(2 marks)**

Question: What exactly did your analysis indicate?
It is perfectly acceptable for your results to be inconclusive. This is a realistic outcome from a statistical piece of coursework.

✦ **The overall piece of work should be of sufficient depth and difficulty.** **(6 marks)**

This section gives credit for all the work that has been done throughout your coursework. Ensure that your approaches have generated enough statistics for you to have shown the appropriate skills from the unit.

You will receive credit not only for the difficulty of your calculations, but also for the quality and depth of your interpretation and for your overall approach.

S2

Your coursework will be assessed at A2 standard so there will be greater expectations of the depth and difficulty of your work than at AS standard.

- **The investigation should form a coherent whole and should be of a length consistent with a piece of work of 8–10 hours. (6 marks)**

Your investigation should read as a logical piece of writing that should be easy to follow. If you struggle to follow the flow of your arguments when reading the work through, then so will the person marking and assessing it.

It is expected that your piece of work will take 8–10 hours to complete. This is an approximate timing and includes data collection and write-up time.

It is important that your piece of work is of sufficient length. Try to design it so that comparisons can be made in your interpretation, but avoid overly long pieces which are 'filled out with waffle'.

Finally, read through your work and check your calculations to ensure that you have not made any careless mistakes.

7.5 Frequently asked questions

Here are some questions that are often asked by students.

- **Should the coursework be hand-written or word-processed?**

It can be either. If it is hand-written try to ensure that the work is neat and clear to follow. If it is word-processed, take care when typing symbols for the sample mean and variance.

- **How long should the piece of work be?**

An appropriate piece of work could vary from 10 sides up to 20 sides including diagrams.

Word-processed pieces tend to be shorter.

- **Should I label the page numbers?**

Yes. It is useful when the work is being moderated.

- **Can I use secondary data?**

Yes. Secondary data is useful for validation. You will be expected to collect your own data for the initial analysis. This could involve sampling from a large database, many of which are available on the Internet.

- **Can I use the Internet?**

Yes. The Internet is appropriate to collect secondary data from as well as other information that may be relevant to your task. Always quote any web-sites used.

Any attempt to copy work from the Internet is against Examination Board rules and could lead to serious consequences.

7.6 Checklist

Have you ...?

Strand 1

- ✦ Stated your aims **clearly**
- ✦ Defined your population **explicitly**
- ✦ Given **details** of how your sample was collected
- ✦ **Discussed** relevant statistical theories **in context**, **embedded** in the work as it arises
- ✦ **Discussed** any assumptions made

Strand 2

- ✦ **Listed** your raw data and organised it appropriatcly
- ✦ **Collected** a large enough amount of data
- ✦ Made **correct** and **appropriate** calculations
- ✦ Used a **sufficient range** of calculations and analysis for your model
- ✦ Used **appropriate** degrees of accuracy in your answers

Strand 3

- ✦ Looked at how **realistic** your final results are, and **if not** have you tried to give reasons
- ✦ Given a **clear statistical** interpretation of your results and made conclusions from them
- ✦ **Related** your results back to the original task **and** discussed if your sampling methods had **any effects** on your results
- ✦ **Discussed** any modifications or improvements that could have been made

Strand 4

- ✦ **Expressed** yourself clearly using **clear** diagrams and appropriate mathematical language and notation
- ✦ **Suggested** other areas of work that could have been considered
- ✦ Set out your **conclusions** clearly
- ✦ Made sure your work is of sufficient **depth and difficulty** (not overly simplistic)
- ✦ Made sure your work is **easy to follow and of sufficient length**

And finally ...

- ✦ **Make sure that you are aware of the deadline set by your teacher and work to it.**

Answers

Unless inappropriate, or requested in the question, or otherwise stated, all decimal answers are given to 3 significant figures.

Chapter 1

Check in

1 A, D **2** A, D **3** a) $\frac{1}{3}$ b) $\frac{1}{18}$ c) $\frac{5}{12}$ **4** $\frac{2}{7}$ **5** 7.6 **6** 8.04

Exercise 1A

1

x	8	9	10
$P(X=x)$	$\frac{1}{3}$	$\frac{1}{2}$	$\frac{1}{6}$

2

x	0	1	2
$P(X=x)$	$\frac{21}{66}$	$\frac{35}{66}$	$\frac{10}{66}$

3

x	0	1	2
$P(X=x)$	$\frac{49}{144}$	$\frac{70}{144}$	$\frac{25}{144}$

4

x	-1	1	4
$P(X=x)$	$\frac{1}{2}$	$\frac{1}{4}$	$\frac{1}{4}$

Exercise 1B

1 $P(X=x)=\frac{1}{10},\ x=0,1,2,\ldots,9$

2

x	0	1
$P(X=x)$	$\frac{5}{6}$	$\frac{1}{6}$

3 $P(X=x)=\frac{1}{2},\ x=0,1$

4

x	0	1	2
$P(x)$	$\frac{1}{4}$	$\frac{1}{2}$	$\frac{1}{4}$

5 B(10, 0.6) **6** X: Not binomial: p is not constant; Y: Not binomial: neither n nor p constant; B(6, 0.19)

7 a) No; b)

x	1	2	3	4	5	…	9
$P(X=x)$	$\frac{111}{300}$	$\frac{111}{300}$	$\frac{12}{300}$	$\frac{11}{300}$	$\frac{11}{300}$	…	$\frac{11}{300}$

Exercise 1C

1 a) 2.5, 7.5 **2** 5, $\frac{77}{3}$ **3** $\frac{29}{9}$, 11 **4** a) $\frac{1}{6}$ b) $\frac{7}{3}$, 6 **5** a) $\frac{1}{14}$, $\frac{18}{7}$, 7

Exercise 1D

1 5, $\frac{19}{6}$, 1.78 **2** 0, $\frac{35}{6}$, 2.42 **3** $\frac{35}{18}$, $\frac{665}{324}$, 1.43 **4** $\frac{5}{4}$, $\frac{45}{112}$, 0.634 **5** $\frac{5}{4}$, $\frac{15}{32}$, 0.685 **6** 2.30, 0.095, 0.308 **7** 0.65, 5.03, 2.24

Exercise 1E

1 a) $\frac{5}{4}$ b) $\frac{13}{4}$ c) $\frac{23}{4}$ **2** a) i) $\frac{9}{4}$ ii) $\frac{33}{4}$ iii) $\frac{5}{4}$ **3** a) i) $\frac{21}{5}$ ii) $\frac{147}{5}$ iii) $-\frac{1}{60}$ **4** a) 2.8, 2.16 b) 1.6, 0.24

5 a) 0.3, 0.06 b) $\frac{1}{2}$, 0

Exercise 1F

1 -4, 9 **2** $\frac{1}{4}$, $\frac{1}{36}$ **3** a) 18 b) 18 c) -6 d) 18 **4** a) 1, 4 b) 10, 36 c) -7, 36 **5** 2, 3 **6** 7.5, 2.5

7 a) 85 b) 81 **8** £50, £2 **9** 10, 50 **10** 1, $\frac{35}{3}$ **11** $(c, d) = (3, 85)$ or $(-3, 115)$

12 $(a, b) = \left(\frac{1}{\sigma}, -\frac{\mu}{\sigma}\right)$ or $\left(\frac{-1}{\sigma}, \frac{\mu}{\sigma}\right)$ **13** σ^2

Check out

1

x	1	2	3	4	5	6	7
$P(X=x)$	$\frac{1}{12}$	$\frac{2}{12}$	$\frac{2}{12}$	$\frac{2}{12}$	$\frac{2}{12}$	$\frac{2}{12}$	$\frac{1}{12}$

2 1.6, 0.84, 0.917 **3** 2.9 **4** 144, 12

5 a) 8.8 b) 26.9 c) 6.7 **6** a) 0.29 b) 1.16

Revision exercise 1

1 a) 3.4 b) 0.84 **2** a) $\frac{1}{3}$ b) i) $\frac{1}{9}$ ii) $\frac{4}{9}$ c) i) $P(X=2)=\frac{1}{9}$ ii) 4, $\frac{4}{3}$ **3** a) 0.6 b) 1.64 **4** 3, $\frac{4}{3}$

5 a) 10, 25 b) 4.8, 6.96 **6** b) 100, 1035 **7** a) 25, 120 b) i)

x	60	30	20	15	12
$P\left(\frac{60}{R}=x\right)$	0.1	0.2	0.4	0.2	0.1

iii) 173.76

8 b) 10, 10.8 c) i) $C = 196 - 4R$ ii) 188, 19.2

9 a) i) 1.47 ii) 3.37 iii) 1.10

9 b) i) B(5, 0.4) ii) 2, 1.10 c) i) 0.20 is very different to 40%

9 ii) Patients in need of treatment will not be a random sample – most patients will be well.

Chapter 2

Check in

1 a) 6 b) 6 c) 1 **2** 4, 8, 16 **3** a) 0.368 b) 0.100 c) 0.0143

Exercise 2A

1 Unlikely: probability of occurrence likely to vary with time (expect higher rate during working days)

2 Unlikely: probability of occurrence likely to vary with time (expect lower rate at night)

3 A reasonable approximation over short periods **4** Will not: traffic not independent of one another

5 Possible: assuming well stirred bun mix **6** Will not: all 6 **7** Will not: range limited to 0–6

8 Possible: during working week **9** Possible: positions may not be independent

10 Unlikely: lions hunt in packs, and live in prides **11** Will not: remnants vary in size

12 Possible: within a single piece of cloth **13** Possible: though rate may depend on tiredness of typist

14 Unlikely: expect rate to vary with page content

Exercise 2B

1 a) 0.741 b) 0.247 c) 0.223 d) 0.135 **2** a) 0.446 b) 2.622

Exercise 2C

1 a) 0.135 b) 0.271 c) 0.271 d) 0.677 e) 0.594 **2** a) 0.986 b) 0.0900 c) 0.0144 **3** a) 0.175 b) 0.440 c) 0.384

4 0.469 **5** 3 **6** a) 0.0824 b) 0.108 c) 0.809 d) 0.918 **7** 0.0341 **8** 0.256 **9** 0.368

Exercise 2D

1 0.135, 0.271, 0.271,0.180, 0.090 **2** 0.175, 0.175, 0.146, 0.104 **3** 0.14, 0.257, 0.205, 0.082

4 a) $x \leqslant \lambda < x + 1$ b) i) 3 ii) 5

Exercise 2E

(Numerical answers to this Exercise are given to 4 d.p.)

1 a) 0.9161 b) 0.9665 **2** a) 0.0629 b) 0.0023 **3** a) 0.1128 b) 0.3359 **4** a) 0.3679 b) 0.1839 **5** 1.8

6 2.0 **7** 0.0418 **8** a) 0.4628 b) 0.1607 **9** a) 0.8335 b) 0.0394 c) 0.4082

10 a) Poisson b) i) 0.5697 ii) 0.1254 c) i) 0.0203 ii) 0.0630

Exercise 2F

1 0.124 **2** 0.498 **3** a) 0.482 b) 0.819 **4** 0.489 **5** 0.395 **6** a) i) 0.378 ii) 0.137 b) 2.74

Check out

1 a) No b) No c) No d) Yes e) Yes **2** a) 0.211 b) 0.042 c) 0.614 **3** a) 0.224 b) 0.050 c) 0.577 **4** 6

5 a) 0.518 b) 0.631 c) 0.809

Revision exercise 2

2 a) 0.673 b) 0.146 **3** a) i) 0.697 ii) 0.339 **4** a) i) 0.0214 ii) 0.136 b) i) 7.5 ii) 0.138

5 a) i) 0.301 ii) 0.217 b) 0.016 c) 0.867 d) Non-independence – may be several passengers in a car

6 a) i) 0.231 ii) 0.531 b) 0.109 c) i) 1.3, 3.34 ii) Variance very different to mean iii) Rate varies with time of day

7 a) i) 0.834 ii) 0.308 b) 0.779 c) 0.091 d) Some people might bring several registered letters at one time

8 a) i) 0.953 ii) 0.551 iii) 0.360 b) i) Poisson, mean 4 ii) 0.371 iii) 0.0506

9 a) 0.977 b) 0.185 c) Non-independence

Chapter 3

Check in

1

value	0–	10–	20–	30–50
frequency	4	5	5	6

3 37.6; 30.0, 46.3 **4** a) $2x$ b) $9x^2 - 2$

5 a) i) 4 ii) $\frac{64}{3}$ iii) $\frac{8}{3}$ b) i) $\frac{1}{2}x^2$ ii) $\frac{1}{3}(x^3 + 1)$ iii) $\frac{1}{4}(-x^4 + 4x^2 - 3)$

Exercise 3A

1 b) $F(x) = \begin{cases} 0 & x \leqslant 1 \\ \frac{1}{16}(x^2 + 4x - 5) & 1 \leqslant x \leqslant 3 \\ 1 & x \geqslant 3 \end{cases}$ **2** b) $F(x) = \begin{cases} 0 & x \leqslant -2 \\ \frac{1}{8}(4 - x^2) & -2 \leqslant x \leqslant 0 \\ \frac{1}{8}(4 + x^2) & 0 \leqslant x \leqslant 2 \\ 1 & x \geqslant 2 \end{cases}$ **3** b) $F(x) = \begin{cases} 0 & x \leqslant 1 \\ \frac{1}{2}(x - 1) & 1 \leqslant x \leqslant 2 \\ \frac{1}{4}x & 2 \leqslant x \leqslant 4 \\ 1 & x \geqslant 4 \end{cases}$

4 b) $F(x) = \begin{cases} 0 & x \leqslant -2 \\ \frac{1}{9}(x^3 + 8) & -2 \leqslant x \leqslant 1 \\ 1 & x \geqslant 1 \end{cases}$ **5** b) $F(x) = \begin{cases} 0 & x \leqslant 0 \\ \frac{1}{19}\{(x + 2)^3 - 8\} & 0 \leqslant x \leqslant 1 \\ 1 & x \geqslant 1 \end{cases}$

6 a) No: graph crosses x-axis b) Yes ($k = \frac{1}{3}$) c) No: area under $y = f(x)$ greater than 1 d) Yes ($k = 0.125$)

7 b) k c) $\frac{3}{4}$ d) $F(x) = \begin{cases} 0 & x \leqslant 1 \\ \frac{1}{4}(x^3 - 3x + 2) & 1 \leqslant x \leqslant 2 \\ 1 & x \geqslant 2 \end{cases}$

Exercise 3B

1 $f(x) = \begin{cases} 1 & 1 < x < 2 \\ 0 & \text{otherwise} \end{cases}$ **2** a) $\frac{1}{16}$ b) $f(x) = \begin{cases} \frac{1}{8}x & 0 < x < 4 \\ 0 & \text{otherwise} \end{cases}$ **3** a) $a = -\frac{1}{8}, b = \frac{1}{8}$ b) $f(x) = \begin{cases} \frac{1}{4}x & 1 < x < 3 \\ 0 & \text{otherwise} \end{cases}$

4 $f(x) = \begin{cases} \frac{1}{4}x & 4 < x < 8 \\ 0 & \text{otherwise} \end{cases}$ **5** a) $a = 1, b = -\frac{1}{8}$ b) $f(x) = \begin{cases} \frac{1}{4}x & 0 < x < 2 \\ \frac{1}{4}(4 - x) & 2 < x < 4 \\ 0 & \text{otherwise} \end{cases}$

Exercise 3C

1 a) $\frac{1}{18}$ b) $\frac{5}{27}$ c) $\frac{14}{27}$ **2** a) $F(x) = \begin{cases} 0 & x \leqslant -2 \\ \frac{1}{3}k(x^3 + 8) & -2 \leqslant x \leqslant 2 \\ 1 & x \geqslant 2 \end{cases}$ b) $\frac{3}{16}$ c) $\frac{7}{16}$ d) $\frac{1}{8}$ e) $\frac{1}{16}$

3 a) $F(x) = \begin{cases} 0 & x \leqslant 0 \\ \frac{1}{4}(x^2 - 1) & 0 \leqslant x \leqslant k \\ 1 & x \geqslant k \end{cases}$ b) $\sqrt{5}$ c) $\frac{7}{16}$ d) 0 e) $\frac{1}{4}$ **4** a) $\frac{1}{2}$ b) $\frac{3}{4}$ **5** a) 1 b) 0 c) $\frac{3}{4}$

6 a) $\frac{1}{24}$ b) 3 c) $\frac{13}{36}$ **7** a) 12 b) $\frac{1}{10000}(600c^2 - 80c^3 + 3c^4)$ c) 0.916, 0.949, 0.973 d) 750 gallons

Exercise 3D

1 b) 2.24, 1.73, 2.65 **2** b) 1, $\frac{1}{2}$, 3 **3** a) $\frac{1}{8}$ b) $F(x) = \begin{cases} 0 & x \leqslant 1 \\ \frac{1}{16}(x^2 + 4x - 5) & 1 \leqslant x \leqslant 3 \\ 1 & x \geqslant 3 \end{cases}$ c) $\frac{9}{16}$ d) $\sqrt{17} - 2$ (= 2.12)

3 e) $\sqrt{13} - 2$ (= 1.61), $\sqrt{21} - 2$ (= 2.58) **4** a) $\frac{1}{4}$ b) 0 c) $F(x) = \begin{cases} 0 & x \leqslant -2 \\ \frac{1}{8}(4 - x^2) & -2 \leqslant x \leqslant 0 \\ \frac{1}{8}(4 + x^2) & 0 \leqslant x \leqslant 2 \\ 1 & x \geqslant 2 \end{cases}$ d) $\frac{1}{8}$ **5** a) $\frac{1}{3}$

5 b) $F(x) = \begin{cases} 0 & x \leqslant -2 \\ \frac{1}{9}(x^3 + 8) & -2 \leqslant x \leqslant 1 \\ 1 & x \geqslant 1 \end{cases}$ c) -1.52 **6** a) $\frac{3}{19}$ b) 0.596 c) 0.393, 0.772 d) $f(x) = \begin{cases} \frac{3}{19}(x + 2)^2 & 0 < x < 1 \\ 0 & \text{otherwise} \end{cases}$

7 a) -6 b) $F(x) = \begin{cases} 0 & x \leqslant 3 \\ (x - 3)^2 & 3 \leqslant x \leqslant 4 \\ 1 & x \geqslant 4 \end{cases}$ c) 3.5, 3.87 **8** a) 1 b) $F(x) = \begin{cases} 0 & x \leqslant 0 \\ 2x - x^2 & 0 \leqslant x \leqslant 1 \\ 1 & x \geqslant 1 \end{cases}$ c) 0.293 d) 0.106, 0.553

9 a) $\frac{2}{7}$ b) $F(s) = \begin{cases} 0 & s \leqslant 0 \\ \frac{1}{7}\{(3+s)^2 - 9\} & 0 \leqslant s \leqslant 1 \\ 1 & s \geqslant 1 \end{cases}$ c) $\frac{15}{28}$ d) 0.536 **10** a) $\frac{1}{129}$ b) $F(t) = \begin{cases} 0 & t \leqslant 5 \\ \frac{1}{387}(t^3 - 125) & 5 \leqslant t \leqslant 8 \\ 1 & t \geqslant 8 \end{cases}$

10 c) 5.25, 7.90 d) 0.461 **11** a) $a = 1.4$ $b = -0.8$ b) 0.404 c) 0.189, 0.660

Exercise 3E

1 a) $\frac{4}{3}$ b) 2 c) $\frac{2}{9}$ d) $\frac{4}{9}$ **2** a) $\frac{13}{6}$ b) $\frac{11}{36}$ **3** a) $\frac{1}{2}$ b) $\frac{3}{2}$ c) $\frac{5}{12}$ d) $\frac{3}{2}$ e) $\frac{1}{2}$ f) $\frac{1}{2}$ **4** a) $\frac{14}{3}$ b) $\frac{2}{3}\sqrt{2}$

5 a) 2 b) 1 c) 0.325 **6** a) 1 b) $\frac{7}{15}$ **7** b) $\frac{1}{3}$ c) 0.5

Exercise 3F

1 a) $f(x) = \begin{cases} \frac{1}{2} & 0 < x < 2 \\ 0 & \text{otherwise} \end{cases}$ $F(x) = \begin{cases} 0 & x \leqslant 0 \\ \frac{1}{2}x & 0 \leqslant x \leqslant 2 \\ 1 & x \geqslant 2 \end{cases}$ **2** a) 1, 7 b) $\frac{1}{3}$ c) $\frac{1}{6}$ **3** a) 4 b) $\frac{1}{3}$ **4** a) 0.4 b) $\frac{1}{\sqrt{3}}$ $(= 0.577)$

5 1, 9 **6** a) 7 b) 3 **7** a) $\frac{1}{8}$ b) $-\frac{15}{4}$ c) $\frac{1}{192}$ d) $\frac{1}{48}$ **8** $\frac{1}{4}$

Check out

1 a) $F(x) = \begin{cases} 0 & x \leqslant 0 \\ \frac{1}{16}\{(x-2)^3 + 8\} & 0 \leqslant x \leqslant 4 \\ 1 & x \geqslant 4 \end{cases}$ b) $f(x) = \begin{cases} \frac{1}{8}(x+1) & -1 < x < 3 \\ 0 & \text{otherwise} \end{cases}$ **2** a) $\frac{1}{16}$ b) $\frac{5}{16}$

3 a) 3.36 b) 0.206, 1.79 c) 0.9 **4** $\frac{26}{9}$, $\frac{26}{81}$, 0.567

5 a) 9, 3 b) i) 0 ii) 16 iii) 160 iv) 16 c) i) $\frac{3}{2}$ ii) 3 iii) $\frac{3}{4}$ iv) 15 v) 12 **6** 2.5 cm, $\frac{25}{12}$ cm^2

Revision exercise 3

1 a) $\frac{1}{90}$ b) $\frac{2}{3}$ c) 185, 675 d) N(185, 9) **2** c) $\frac{23}{32}$ **3** b) i) $\frac{4}{9}$ ii) $\frac{1}{4}$ **4** a) 0.3 b) ii) 0.125 **5** a) $a + \frac{1}{2}k, \frac{1}{12}k^2$ c) $\frac{2}{3}$

6 b) $F(t) = \begin{cases} 0 & t \leqslant 0 \\ \frac{1}{54}t^3 & 0 \leqslant t \leqslant 3 \\ \frac{1}{8}(10t - t^2 - 17) & 3 \leqslant t \leqslant 5 \\ 1 & t \geqslant 5 \end{cases}$ c) $\frac{7}{8}$ **7** a) ii) 5.45 b) $\frac{213}{40}$ **8** b) i) $\frac{1}{2}$ ii) $\frac{1}{3}$ iii) 3.5 iv) 4.94%

9 b) i) 1 ii) $\sqrt{2}$ c) A is not valid $(2c + 2d \neq 1)$; B is valid; C is not valid ($f(x) < 0$ for $0 < x < \frac{2}{3}$)

Chapter 4

Check in

1 14, 24.8 **2** N(13, 3.2) **3** N(3, 0.5)

Exercise 4A

Answers to **1** are given to the tabular accuracy of 3 dp.

1 a) 1.860 b) 2.602 c) 2.571 d) −2.681 e) −2.776 f) 1.812 **2** a) 0.01 b) 0.05 c) 0.875

3 a) 0.975 b) 0.995 c) 0.9

Exercise 4B

1 a) (7.08, 9.58) b) (6.53, 10.13) **2** a) (999.56, 1004.81) ml b) 2.38 ml **3** Normal, (1.54, 3.20) m

4 (96.7, 108.1) grams **5** (88.2, 162.0) grams **6** a) 88.8, 36.9, (82.5, 95.0) mph b) 0.421

7 (58.4, 68.4) cm, random sample, normal distribution

Check out

1 a) 3.143 b) 1.782 **2** (6.21, 16.59)

Revision exercise 4

1 (75.66, 76.23) cl **2** a) 40.5 m, 1.91m^2 b) i) (39.3, 41.7) m ii) Random sample, normal distribution

3 a) 95.5 min, 28.2 (min)^2 b) (92.4, 98.6) min **4** a) To avoid bias b) 52.6 grams, 9.53 (grams)^2 c) (50.6, 54.5) grams

4 d) a useful way, but likely to give more than 50 grams

5 a) (454.2, 459.3) grams b) Mean content likely to be more than 454 grams c) 0.05

6 a) (169, 1085) ml b) i) model cannot account for non-drinkers (the 3 zeros) ii) Impossible to consume negative amount

6 c) i) (926.0, 994.0) ml ii) CLT, large sample iii) Data only for night shift

7 a) i) (490.0, 502.2) grams ii) (492.5, 499.8) grams b) (498.0, 501.2) grams c) i) a) 0.8 b) 0.15 ii) 0.19

8 a) (29.80, 67.00) b) i) z: (46.42, 52.58) or t: (46.38, 52.62) ii) z: (17.16, 81.84) or t: (16.80, 82.20) c) Most will achieve at least 25

8 d) Confidence interval narrower, assumption of normality not required, better estimates of μ and σ^2

8 e) Sample of 120 preferable to one of 110

9 a) (135.25, 135.95) mm b) (134.8, 136.8) mm c) i) Likely to decrease

9 c) ii) Likely to increase, since, unless the mean is 135, the narrower interval is less likely to include that value.

9 d) When the sample variance is large, there is a wide confidence interval that may include 135, even though production is unsatisfactory; conversely, if the sample variance is low, then the narrow confidence interval may exclude 135 even though production is satisfactory.

Chapter 5

Check in

In **2** answers are given to full tabular accuracy

1 $P(Z > 0.5)$ **2** a) 1.6449 b) 1.9600 c) −2.5758 d) 1.6449 **3** a) 5.5 b) 505.4 c) 43.275

4 a) 4.425 b) 81 c) 445.54

Exercise 5A

1 $H_0 : \mu = 1, H_1 : \mu \neq 1$ **2** $H_0 : \mu = 41.0, H_1 : \mu < 41.0$ **3** $H_0 : \mu = 1, H_1 : \mu \neq 1$ **4** $H_0 : \mu = 51.2, H_1 : \mu > 51.2$

5 $H_0 : \mu = 2.4, H_1 : \mu \neq 2.4$

Exercise 5B

1 a) Accepting mean weight, μ, $\neq$ 1 kg when μ = 1 kg b) Accepting mean weight, μ, = 1 kg when $\mu \neq$ 1 kg

2 a) Accepting average speed, μ,< 41.0 mph when μ = 41.0 mph b) Accepting average speed, μ, = 41.0 mph when μ < 41.0 mph

3 a) Accepting average weight, μ, $\neq$ 1 gram when μ = 1 gram b) Accepting average weight, μ, = 1 gram when $\mu \neq$ 1 gram

4 a) Accepting average mark improved when it has not b) Accepting average mark unchanged when it has improved

5 a) Accepting average annual crop (per plant), μ, $\neq$ 2.4 kg when μ = 2.4 kg

5 b) Accepting average crop per plant, μ, = 2.4 kg when $\mu \neq$ 2.4 kg

Exercise 5C

1 $H_0 : \mu = 460.3, H_1 : \mu \neq 460.3$; $|z| = 2.18 > 1.9600$ ($p = 0.0294 < 0.05$): Significant evidence that mean has changed

2 $z = -2.71 < -1.6449$ ($p = 0.0034 < 0.05$): Significant evidence that mean time has decreased

3 a) $\bar{x} > 97.67$ b) 0.01 c) Accepting mean life is unchanged when it has increased

3 d) $z = 1.32 < 2.3263$ ($p = 0.0936 > 0.01$): No significant evidence that mean life has increased

4 $z = -1.51 > -1.8808$ ($p = 0.0651 > 0.03$): No significant evidence that value of μ is overstated

5 $|z| = 1.48 < 1.6449$ ($p = 0.140 > 0.10$): No significant evidence that mean is reduced

6 a) $N\left(\mu, \left(\frac{\sigma}{10}\right)^2\right)$ b) $z = 2.16 < 2.3263$ ($p = 0.0154 > 0.01$): No significant evidence that $\mu > 10.5$ (one tailed test)

6 c) i) Accepting $\mu > 10.5$ when $\mu = 10.5$ ii) Accepting $\mu = 10.5$ when $\mu > 10.5$

Exercise 5D

1 $|t| = 0.390 < 2.262$ ($p = 0.706 > 0.05$): No significant evidence that allegation is unreasonable

2 $t = 2.17 > 1.895$ ($p = 0.0334 < 0.05$): Significant evidence that mean weight exceeds 2000 g

3 $|t| = 1.92 < 2.262$ ($p = 0.0875 > 0.05$): No significant evidence that average reaction time is not equal to 5.0 seconds

4 a) $t > 1.796$ b) Accepting mean lifetime, μ, > 1200 hours when μ = 1200 hours

4 c) Accepting mean lifetime, μ, = 1200 hours when μ > 1200 hours d) $t = 1.11 < 1.796$ ($p = 0.145 > 0.05$): No significant evidence that mean lifetime is less than 1200 hours

4 e) Because the value of t will be negative and therefore not in the upper tail of the relevant t-distribution

5 $t = 3.17 > 2.998$ ($p = 0.00791 < 0.01$): Significant evidence that mean yield is higher

6 1.3% a) $|z| = 2.93 > 1.9600$ ($p = 0.00343 < 0.05$): Significant evidence that mean has changed

6 b) $|t| = 1.60 < 2.306$ ($p = 0.149 > 0.05$): No significant evidence that mean has changed

Exercise 5E

1 $z = -1.88 < -1.6449$ ($p = 0.0298 < 0.05$): Significant evidence that journey now takes less time

2 $z = 2.17 < 2.3263$ ($p = 0.0149 > 0.01$): No significant evidence that supplements increase IQ scores

3 $|z| = 3.05 > 2.5758$ ($p = 0.0023 < 0.01$): Significant evidence that mean weight is not 70 grams

4 $|z| = 1.53 < 1.6449$ ($p = 0.126 > 0.10$): No significant evidence that rumour is untrue

5 $z = -2.76 < -1.6449$ ($p = 0.00285 < 0.05$): Significant evidence that mean shopping time is less than 20 minutes

6 $|z| = 1.33 < 1.9600$ ($p = 0.1846 > 0.05$): No significant evidence that mean height is not 1.5 m

Exercise 5F

1 $H_0 : \mu = 460.3, H_1 : \mu \neq 460.3$; (460.4, 462.0): Significant evidence that mean has changed

2 $H_0 : \mu = 5.82, H_1 : \mu < 5.82$; (5.58, 5.78): Significant evidence that tests are carried out more quickly

3 $H_0 : \mu = 95.2, H_1 : \mu \neq 95.2$; (90.9, 96.3): No significant evidence of change in mean life

Check out

1 a) $H_0 : \mu = 250, H_1 : \mu > 250$ b) One tailed c) z d) $z > 1.6449$, 1.6449 e) Type I: Accepting $\mu > 250$ when $\mu = 250$; Type II: Accepting $\mu = 250$ when $\mu > 250$ f) $z = 2 > 1.6449$: Significant evidence that μ has increased from 250

2 $|z| = 1.2 < 1.6449$ ($p = 0.230 > 0.10$): No significant evidence that $\mu \neq 80$

3 $t = 3.00 > 2.602$ ($p = 0.0045 < 0.01$): Significant evidence that $\mu > 20$

4 $z = 1.61 < 1.9600$ ($p = 0.107 > 0.05$): No significant evidence that $\mu \neq 250$ **5** No significant evidence to support H_1

Revision exercise 5

1 $|z| = 2.02 > 1.9600$ ($p = 0.0433 < 0.05$): Significant evidence that mean length has changed from 56 cm

2 a) i) N(40, 0.0128) ii) 0.05 b) (39.78, 40.22) c) Accepting $\mu = 40$ when $\mu \neq 40$

3 a) $H_0 : \mu = 500, H_1 : \mu < 500$ b) N(500, 0.64) c) i) 0.05 ii) $\overline{Y} < 498.68$

4 $|z| = 2.45 > 1.6449$ ($p = 0.014 < 0.1$): Significant evidence that Bill's claim is incorrect

5 $|t| = 1.28 < 1.895$ ($p = 0.242 > 0.1$): No significant evidence that mean weight is not 17 grams

6 a) $H_0 : \mu = 632, H_1 : \mu \neq 632$, $|t| = 0.160 < 2.306$ ($p = 0.877 > 0.05$): No significant evidence of a change in mean payment

6 b) Sample random, normal distribution c) $z = 6.10 > 2.3263$ ($p < 0.01$): Significant evidence of increase in mean payment

6 d) Large sample (CLT) stated to be random

6 e) $\bar{x} = 600.9 < 632$, gives value in lower tail, whereas the critical region is in the upper tail

7 a) $H_0 : \mu = 0, H_1 : \mu \neq 0$, $|t| = 0.663 < 2.365$ ($p = 0.529 > 0.05$): No significant evidence that instrument is biased

7 b) i) $z = -2.94 < -2.3263$ ($p = 0.0016 < 0.01$): Significant evidence that second instrument underestimates speed

7 b) ii) May be preferable to have a smaller standard deviation

8 a) $H_0 : \mu = 50, H_1 : \mu \neq 50$, $|t| = 2.42 > 2.365$ ($p = 0.046 < 0.05$): Significant evidence that mean mark differs from 50

8 b) Random sample, normal population c) i) $H_0 : \mu = 48.8, H_1 : \mu > 48.8, z = 1.20 < 2.3263$ ($p = 0.115 > 0.01$): No significant evidence that mean mark is greater than 48.8 ii) Sample is self-selected (students accepting book offer – likely, therefore to be keen!) and not random

9 a) $t = -1.66 > -1.796$ ($p = 0.062 > 0.05$): No significant evidence that mean lifetime is less than 40 months

9 b) $z = -2.41 < -1.6499$ ($p = 0.0080 < 0.05$): Significant evidence that mean lifetime is less than 40 months

9 c) i) The Type I error is 5% in both cases

Chapter 6

Check in

1 a) $\frac{1}{12}$ b) $\frac{1}{6}$ **2** a) 5 b) $\frac{25}{3}$

Exercise 6A

1 a) 0.95 b) 0.99 c) 0.99 **2** a) 0.025 b) 0.01 c) 0.1 **3** a) 14.860 b) 3.841 c) 55.758 d) 10.597

4 a) 0.045 b) 0.05

Exercise 6B

1 $\chi^2 = 7.55 < 9.488$ ($p = 0.110 > 0.05$): No significant evidence of difference in effectiveness

2 $\chi^2 = 19.3 > 13.277$ ($p = 0.00069 < 0.01$): Significant evidence of association between greatest academic achievement and party supported

S2

3 (Combine 16–17 with 18–24) $\chi^2 = 3.47 < 7.815$ ($p = 0.324 > 0.05$): No significant evidence of difference in age distribution of males and females

4 a) Table gives percentages, not frequencies b) H_0: Proportion passing is independent of driving school

4 b) $\chi^2 = 4.98 < 5.991$ ($p = 0.083 > 0.05$): No significant evidence that proportion passing is dependent on driving school

5 a) To avoid expected values < 5 b) $\chi^2 = 7.19 < 7.779$ ($p = 0.126 > 0.10$): No significant evidence of association between A-level grade and degree class

6 a) $\chi^2 = 17.0 > 9.210$ ($p = 0.000203 < 0.05$): Significant evidence that proportion transferred is not independent of surgeon

6 b) i) To reduce proportion transferred

6 b) ii) The high proportion of C's cases that were transferred might be a consequence of his having been given the most difficult cases: need more detail about the cases

7 a)

	X	Y	Z
London	42	63	20
Birmingham	21	41	28

b) $\chi^2 = 7.49 > 5.991$ ($p = 0.0237 < 0.05$): Significant evidence of different preferences

Exercise 6C

1 $\chi_c^2 = 4.01 > 3.841$: Significant evidence of association between sex of driver and age of car

2 a)

	bad health	moderate health	excellent health
for	276	232	73
against	41	26	13

2 b) $\chi^2 = 1.94 < 4.605$ ($p = 0.0380 < 0.1$): No significant evidence that attitude to early retirement depends on health

2 c) $\chi_c^2 = 0.121, 2.706$: No significant evidence that attitude to early retirement depends on health

3 $\chi_c^2 = 1.06 > 3.841$: No significant evidence that proportions of librarians and statisticians are unequal

Check out

1 a) 16.919 b) 13.277 c) 2.706

2 $\chi^2 = 4.40 < 9.488$ ($p = 0.355 < 0.05$): No significant evidence of dependence between rows and columns

3 $\chi_c^2 = 6.72 > 6.635$: Significant evidence of dependence between rows and columns

Revision exercise 6

1 H_0: test results are independent of driving school; $\chi_c^2 = 3.43 < 3.841$: No significant evidence of association between test results and driving school

2 H_0: length of stay in prison is independent of the frequency with which mail is received,
H_1: length of stay in prison is dependent on the frequency with which mail is received; $\chi^2 = 22.9 > 13.277$ ($p = 0.00013 < 0.01$): Significant evidence of association between length of stay in prison and the frequency with which mail is received

3 H_0: speed of vehicle is independent of the gender of the driver, H_1: speed of vehicle is dependent on the gender of the driver; $\chi^2 = 25.0 > 9.210$ ($p < 0.01$): Significant evidence of association between speed of vehicle and the gender of the driver

4 H_0: number of A, B, C and D grades achieved is independent of school; $\chi^2 = 5.71 < 7.815$ ($p = 0.127 > 0.05$): No significant evidence of association between number of A, B, C and D grades and school

5 a) $\chi^2 = 7.69 > 5.991$ ($p = 0.0214 < 0.05$): Significant evidence of association between choice of language and gender

5 b) i) and ii) Of those who study Japanese and Russian the genders are reasonably balanced, but for Swedish over 70% are male; Swedish is the most popular choice for both males and females

6 a) i) $\chi_c^2 = 6.35 > 3.841$

6 a) ii) Significant evidence of association between whether a passenger carried a mobile phone and whether the passenger is irritated by other passengers using mobile phones

6 b) i) Combining very dissimilar classes makes interpretation difficult

6 b) ii) Anthony's suggestion would retain as much information as possible, whereas Eric's sugggestion would lose more information

6 b) iii) $\chi^2 = 32.3 > 9.210$ ($p << 0.01$): Significant evidence of association between whether a passenger carried a mobile phone and whether the passenger is irritated by other passengers using mobile phones

7 a) $\chi^2 = 23.4 > 9.210$ ($p << 0.01$): Significant evidence of association between whether a customer brings alcohol, or not, and time of day

7 b) i) Between 6pm and 9pm, customers are more likely to bring alcohol ii) There are more customers between 6pm and 9pm than in either of the other time-periods

7 c) Necessary to know that it was a randomly chosen day

8 a) $\chi_c^2 - 8.13 > 3.841$: Significant evidence that reason for using the crossing is associated with time of day

8 b) Those using the crossing because of the Beatles are more likely to use it out of rush hour, and those using the crossing for normal reasons are more likely to use it in the rush hour

9 a) $\chi^2 = 67.3 > 9.210$ ($p << 0.01$) b) Significant evidence of association between area and advertised price

9 c) i) A good idea to collect more detailed data ii) Relationship appears to be non-linear so pmcc unsuitable

MS2A Practice Paper

1 a) 1.32, 0.008 b) (1.23, 1.41) **2** a) $\frac{11}{3}, \frac{14}{9}$ b) $18\pi, 56\pi^2$

3 a) $z = -1.49 > 1.6449$ ($p = 0.0686 > 0.05$: No significant evidence that mean weight is below specification

3 b) Accepting mean weight is not below specification when, in fact, it is **4** a) $\begin{matrix} 7 & 3 \\ 70 & 30 \\ 63 & 27 \\ 35 & 15 \end{matrix}$ b) $E_{1,2} < 5$

4 c) $\chi^2 = 13.9 > 5.991$ ($p = 0.000965 < 0.05$) : Significant evidence of association between number of shares held and view on proposal

4 d) Smaller shareholders tend to be in favour of proposal and larger shareholders tend to be against proposal

5 a) i) 0.663 ii) 0.125 b) 1.116 **6** b) 0 c) $\frac{5}{16}$ d) i) Symmetry of distribution ii) $\frac{1}{20}$ e) It is not impossible that $|x| > 0.5$

6 f) reduction of range (or variance or standard deviation)

MS2B Practice Paper

1 a) 1.32, 0.008 b) (1.23, 1.41) **2** a) $\frac{11}{3}, \frac{14}{9}$ b) $18\pi, 56\pi^2$ **3** a) $c, 3c^2$ c) $\frac{1}{2}$ **4** a) $\begin{matrix} 7 & 3 \\ 70 & 30 \\ 63 & 27 \\ 35 & 15 \end{matrix}$ b) $E_{1,2} < 5$

4 c) $\chi^2 = 13.9 > 5.991$ ($p = 0.000965 < 0.05$) : Significant evidence of association between number of shares held and view on proposal

4 d) Smaller shareholders tend to be in favour of proposal and larger shareholders tend to be against proposal

5 a) i) 0.394 ii) 0.0476 b) 0.105 c) 0.0331 **6** b) 0 c) $\frac{5}{16}$ d) i) Symmetry of distribution ii) $\frac{1}{20}$

6 e) It is not impossible that $|x| > 0.5$ f) reduction of range (or variance or standard deviation)

7 a) H_0 : mean viscosity = 34.2 pascal seconds, H_1 : mean viscosity is greater than 34.2 pascal seconds

7 b) Probability of accepting that the oil does not meet the spcification when, in fact, it does (that is, accepting mean viscosity is greater than the specified value of 34.2 pascal seconds when, in fact it is equal to 34.2 seconds) is 0.05

7 c) Individual observations are independent and have a normal distribution d) $k = 34.53$

7 e) Set of values of sample mean that lead to accepting the claim that the viscosity exceeds specification

7 f) $34.45 < 34.53$: Accept that oil meets the specification, that is, mean viscosity is 34.2 pascal seconds

7 g) Accepting oil meets specification that mean viscosity is 34.2 pascal seconds, when the mean viscosity is greater than 34.2 pascal seconds

Appendices

CUMULATIVE POISSON DISTRIBUTION FUNCTION (Table 2)

The tabulated value is $P(X \leqslant x)$, where X has a Poisson distribution with mean λ.

λ	0.10	0.20	0.30	0.40	0.50	0.60	0.70	0.80	0.90	1.0	1.2	1.4	1.6	1.8	λ
x															x
0	0.9048	0.8187	0.7408	0.6703	0.6065	0.5488	0.4966	0.4493	0.4066	0.3679	0.3012	0.2466	0.2019	0.1653	0
1	0.9953	0.9825	0.9631	0.9384	0.9098	0.8781	0.8442	0.8088	0.7725	0.7358	0.6626	0.5918	0.5249	0.4628	1
2	0.9998	0.9989	0.9964	0.9921	0.9856	0.9769	0.9659	0.9526	0.9371	0.9197	0.8795	0.8335	0.7834	0.7306	2
3	1.0000	0.9999	0.9997	0.9992	0.9982	0.9966	0.9942	0.9909	0.9865	0.9810	0.9662	0.9463	0.9212	0.8913	3
4		1.0000	1.0000	0.9999	0.9998	0.9996	0.9992	0.9986	0.9977	0.9963	0.9923	0.9857	0.9763	0.9636	4
5				1.0000	1.0000	1.0000	0.9999	0.9998	0.9997	0.9994	0.9985	0.9968	0.9940	0.9896	5
6							1.0000	1.0000	1.0000	0.9999	0.9997	0.9994	0.9987	0.9974	6
7										1.0000	1.0000	0.9999	0.9997	0.9994	7
8												1.0000	1.0000	0.9999	8
9														1.0000	9

λ	2.0	2.2	2.4	2.6	2.8	3.0	3.2	3.4	3.6	3.8	4.0	4.5	5.0	5.5	λ
x															x
0	0.1353	0.1108	0.0907	0.0743	0.0608	0.0498	0.0408	0.0334	0.0273	0.0224	0.0183	0.0111	0.0067	0.0041	0
1	0.4060	0.3546	0.3084	0.2674	0.2311	0.1991	0.1712	0.1468	0.1257	0.1074	0.0916	0.0611	0.0404	0.0266	1
2	0.6767	0.6227	0.5697	0.5184	0.4695	0.4232	0.3799	0.3397	0.3027	0.2689	0.2381	0.1736	0.1247	0.0884	2
3	0.8571	0.8194	0.7787	0.7360	0.6919	0.6472	0.6025	0.5584	0.5152	0.4735	0.4335	0.3423	0.2650	0.2017	3
4	0.9473	0.9275	0.9041	0.8774	0.8477	0.8153	0.7806	0.7442	0.7064	0.6678	0.6288	0.5321	0.4405	0.3575	4
5	0.9834	0.9751	0.9643	0.9510	0.9349	0.9161	0.8946	0.8705	0.8441	0.8156	0.7851	0.7029	0.6160	0.5289	5
6	0.9955	0.9925	0.9884	0.9828	0.9756	0.9665	0.9554	0.9421	0.9267	0.9091	0.8893	0.8311	0.7622	0.6860	6
7	0.9989	0.9980	0.9967	0.9947	0.9919	0.9881	0.9832	0.9769	0.9692	0.9599	0.9489	0.9134	0.8666	0.8095	7
8	0.9998	0.9995	0.9991	0.9985	0.9976	0.9962	0.9943	0.9917	0.9883	0.9840	0.9786	0.9597	0.9319	0.8944	8
9	1.0000	0.9999	0.9998	0.9996	0.9993	0.9989	0.9982	0.9973	0.9960	0.9942	0.9919	0.9829	0.9682	0.9462	9
10		1.0000	1.0000	0.9999	0.9998	0.9997	0.9995	0.9992	0.9987	0.9981	0.9972	0.9933	0.9863	0.9747	10
11				1.0000	1.0000	0.9999	0.9999	0.9998	0.9996	0.9994	0.9991	0.9976	0.9945	0.9890	11
12						1.0000	1.0000	0.9999	0.9999	0.9998	0.9997	0.9992	0.9980	0.9955	12
13								1.0000	1.0000	1.0000	0.9999	0.9997	0.9993	0.9983	13
14											1.0000	0.9999	0.9998	0.9994	14
15												1.0000	0.9999	0.9998	15
16													1.0000	0.9999	16
17														1.0000	17

λ	6.0	6.5	7.0	7.5	8.0	8.5	9.0	9.5	10.0	11.0	12.0	13.0	14.0	15.0	λ
x															*x*
0	0.0025	0.0015	0.0009	0.0006	0.0003	0.0002	0.0001	0.0001	0.0000	0.0000	0.0000	0.0000	0.0000	0.0000	**0**
1	0.0174	0.0113	0.0073	0.0047	0.0030	0.0019	0.0012	0.0008	0.0005	0.0002	0.0001	0.0000	0.0000	0.0000	**1**
2	0.0620	0.0430	0.0296	0.0203	0.0138	0.0093	0.0062	0.0042	0.0028	0.0012	0.0005	0.0002	0.0001	0.0000	**2**
3	0.1512	0.1118	0.0818	0.0591	0.0424	0.0301	0.0212	0.0149	0.0103	0.0049	0.0023	0.0011	0.0005	0.0002	**3**
4	0.2851	0.2237	0.1730	0.1321	0.0996	0.0744	0.0550	0.0403	0.0293	0.0151	0.0076	0.0037	0.0018	0.0009	**4**
5	0.4457	0.3690	0.3007	0.2414	0.1912	0.1496	0.1157	0.0885	0.0671	0.0375	0.0203	0.0107	0.0055	0.0028	**5**
6	0.6063	0.5265	0.4497	0.3782	0.3134	0.2562	0.2068	0.1649	0.1301	0.0786	0.0458	0.0259	0.0142	0.0076	**6**
7	0.7440	0.6728	0.5987	0.5246	0.4530	0.3856	0.3239	0.2687	0.2202	0.1432	0.0895	0.0540	0.0316	0.0180	**7**
8	0.8472	0.7916	0.7291	0.6620	0.5925	0.5231	0.4557	0.3918	0.3328	0.2320	0.1550	0.0998	0.0621	0.0374	**8**
9	0.9161	0.8774	0.8305	0.7764	0.7166	0.6530	0.5874	0.5218	0.4579	0.3405	0.2424	0.1658	0.1094	0.0699	**9**
10	0.9574	0.9332	0.9015	0.8622	0.8159	0.7634	0.7060	0.6453	0.5830	0.4599	0.3472	0.2517	0.1757	0.1185	**10**
11	0.9799	0.9661	0.9467	0.9208	0.8881	0.8487	0.8030	0.7520	0.6968	0.5793	0.4616	0.3532	0.2600	0.1848	**11**
12	0.9912	0.9840	0.9730	0.9573	0.9362	0.9091	0.8758	0.8364	0.7916	0.6887	0.5760	0.4631	0.3585	0.2676	**12**
13	0.9964	0.9929	0.9872	0.9784	0.9658	0.9486	0.9261	0.8981	0.8645	0.7813	0.6815	0.5730	0.4644	0.3632	**13**
14	0.9986	0.9970	0.9943	0.9897	0.9827	0.9726	0.9585	0.9400	0.9165	0.8540	0.7720	0.6751	0.5704	0.4657	**14**
15	0.9995	0.9988	0.9976	0.9954	0.9918	0.9862	0.9780	0.9665	0.9513	0.9074	0.8444	0.7636	0.6694	0.5681	**15**
16	0.9998	0.9996	0.9990	0.9980	0.9963	0.9934	0.9889	0.9823	0.9730	0.9441	0.8987	0.8355	0.7559	0.6641	**16**
17	0.9999	0.9998	0.9996	0.9992	0.9984	0.9970	0.9947	0.9911	0.9857	0.9678	0.9370	0.8905	0.8272	0.7489	**17**
18	1.0000	0.9999	0.9999	0.9997	0.9993	0.9987	0.9976	0.9957	0.9928	0.9823	0.9626	0.9302	0.8826	0.8195	**18**
19		1.0000	1.0000	0.9999	0.9997	0.9995	0.9989	0.9980	0.9965	0.9907	0.9787	0.9573	0.9235	0.8752	**19**
20				1.0000	0.9999	0.9998	0.9996	0.9991	0.9984	0.9953	0.9884	0.9750	0.9521	0.9170	**20**
21					1.0000	0.9999	0.9998	0.9996	0.9993	0.9977	0.9939	0.9859	0.9712	0.9469	**21**
22						1.0000	0.9999	0.9999	0.9997	0.9990	0.9970	0.9924	0.9833	0.9673	**22**
23							1.0000	0.9999	0.9999	0.9995	0.9985	0.9960	0.9907	0.9805	**23**
24								1.0000	1.0000	0.9998	0.9993	0.9980	0.9950	0.9888	**24**
25										0.9999	0.9997	0.9990	0.9974	0.9938	**25**
26										1.0000	0.9999	0.9995	0.9987	0.9967	**26**
27											0.9999	0.9998	0.9994	0.9983	**27**
28											1.0000	0.9999	0.9997	0.9991	**28**
29												1.0000	0.9999	0.9996	**29**
30													0.9999	0.9998	**30**
31													1.0000	0.9999	**31**
32														1.0000	**32**

NORMAL DISTRIBUTION FUNCTION (Table 3)

The table gives the probability, p, that a normally distributed random variable Z, with mean = 0 and variance = 1, is less than or equal to z.

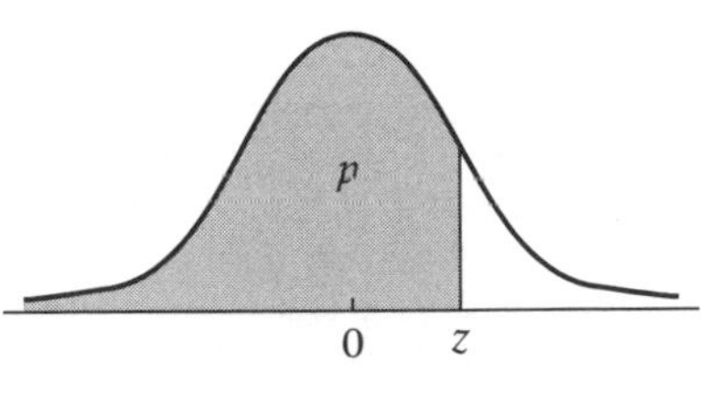

z	0.00	0.01	0.02	0.03	0.04	0.05	0.06	0.07	0.08	0.09	z
0.0	0.50000	0.50399	0.50798	0.51197	0.51595	0.51994	0.52392	0.52790	0.53188	0.53586	**0.0**
0.1	0.53983	0.54380	0.54776	0.55172	0.55567	0.55962	0.56356	0.56749	0.57142	0.57535	**0.1**
0.2	0.57926	0.58317	0.58706	0.59095	0.59483	0.59871	0.60257	0.60642	0.61026	0.61409	**0.2**
0.3	0.61791	0.62172	0.62552	0.62930	0.63307	0.63683	0.64058	0.64431	0.64803	0.65173	**0.3**
0.4	0.65542	0.65910	0.66276	0.66640	0.67003	0.67364	0.67724	0.68082	0.68439	0.68793	**0.4**
0.5	0.69146	0.69497	0.69847	0.70194	0.70540	0.70884	0.71226	0.71566	0.71904	0.72240	**0.5**
0.6	0.72575	0.72907	0.73237	0.73565	0.73891	0.74215	0.74537	0.74857	0.75175	0.75490	**0.6**
0.7	0.75804	0.76115	0.76424	0.76730	0.77035	0.77337	0.77637	0.77935	0.78230	0.78524	**0.7**
0.8	0.78814	0.79103	0.79389	0.79673	0.79955	0.80234	0.80511	0.80785	0.81057	0.81327	**0.8**
0.9	0.81594	0.81859	0.82121	0.82381	0.82639	0.82894	0.83147	0.83398	0.83646	0.83891	**0.9**
1.0	0.84134	0.84375	0.84614	0.84849	0.85083	0.85314	0.85543	0.85769	0.85993	0.86214	**1.0**
1.1	0.86433	0.86650	0.86864	0.87076	0.87286	0.87493	0.87698	0.87900	0.88100	0.88298	**1.1**
1.2	0.88493	0.88686	0.88877	0.89065	0.89251	0.89435	0.89617	0.89796	0.89973	0.90147	**1.2**
1.3	0.90320	0.90490	0.90658	0.90824	0.90988	0.91149	0.91309	0.91466	0.91621	0.91774	**1.3**
1.4	0.91924	0.92073	0.92220	0.92364	0.92507	0.92647	0.92785	0.92922	0.93056	0.93189	**1.4**
1.5	0.93319	0.93448	0.93574	0.93699	0.93822	0.93943	0.94062	0.94179	0.94295	0.94408	**1.5**
1.6	0.94520	0.94630	0.94738	0.94845	0.94950	0.95053	0.95154	0.95254	0.95352	0.95449	**1.6**
1.7	0.95543	0.95637	0.95728	0.95818	0.95907	0.95994	0.96080	0.96164	0.96246	0.96327	**1.7**
1.8	0.96407	0.96485	0.96562	0.96638	0.96712	0.96784	0.96856	0.96926	0.96995	0.97062	**1.8**
1.9	0.97128	0.97193	0.97257	0.97320	0.97381	0.97441	0.97500	0.97558	0.97615	0.97670	**1.9**
2.0	0.97725	0.97778	0.97831	0.97882	0.97932	0.97982	0.98030	0.98077	0.98124	0.98169	**2.0**
2.1	0.98214	0.98257	0.98300	0.98341	0.98382	0.98422	0.98461	0.98500	0.98537	0.98574	**2.1**
2.2	0.98610	0.98645	0.98679	0.98713	0.98745	0.98778	0.98809	0.98840	0.98870	0.98899	**2.2**
2.3	0.98928	0.98956	0.98983	0.99010	0.99036	0.99061	0.99086	0.99111	0.99134	0.99158	**2.3**
2.4	0.99180	0.99202	0.99224	0.99245	0.99266	0.99286	0.99305	0.99324	0.99343	0.99361	**2.4**
2.5	0.99379	0.99396	0.99413	0.99430	0.99446	0.99461	0.99477	0.99492	0.99506	0.99520	**2.5**
2.6	0.99534	0.99547	0.99560	0.99573	0.99585	0.99598	0.99609	0.99621	0.99632	0.99643	**2.6**
2.7	0.99653	0.99664	0.99674	0.99683	0.99693	0.99702	0.99711	0.99720	0.99728	0.99736	**2.7**
2.8	0.99744	0.99752	0.99760	0.99767	0.99774	0.99781	0.99788	0.99795	0.99801	0.99807	**2.8**
2.9	0.99813	0.99819	0.99825	0.99831	0.99836	0.99841	0.99846	0.99851	0.99856	0.99861	**2.9**
3.0	0.99865	0.99869	0.99874	0.99878	0.99882	0.99886	0.99889	0.99893	0.99896	0.99900	**3.0**
3.1	0.99903	0.99906	0.99910	0.99913	0.99916	0.99918	0.99921	0.99924	0.99926	0.99929	**3.1**
3.2	0.99931	0.99934	0.99936	0.99938	0.99940	0.99942	0.99944	0.99946	0.99948	0.99950	**3.2**
3.3	0.99952	0.99953	0.99955	0.99957	0.99958	0.99960	0.99961	0.99962	0.99964	0.99965	**3.3**
3.4	0.99966	0.99968	0.99969	0.99970	0.99971	0.99972	0.99973	0.99974	0.99975	0.99976	**3.4**
3.5	0.99977	0.99978	0.99978	0.99979	0.99980	0.99981	0.99981	0.99982	0.99983	0.99983	**3.5**
3.6	0.99984	0.99985	0.99985	0.99986	0.99986	0.99987	0.99987	0.99988	0.99988	0.99989	**3.6**
3.7	0.99989	0.99990	0.99990	0.99990	0.99991	0.99991	0.99992	0.99992	0.99992	0.99992	**3.7**
3.8	0.99993	0.99993	0.99993	0.99994	0.99994	0.99994	0.99994	0.99995	0.99995	0.99995	**3.8**
3.9	0.99995	0.99995	0.99996	0.99996	0.99996	0.99996	0.99996	0.99996	0.99997	0.99997	**3.9**

PERCENTAGE POINTS OF THE NORMAL DISTRIBUTION (Table 4)

The table gives the values of z satisfying $P(Z \leqslant z) = p$, where Z is the normally distributed random variable with mean = 0 and variance = 1.

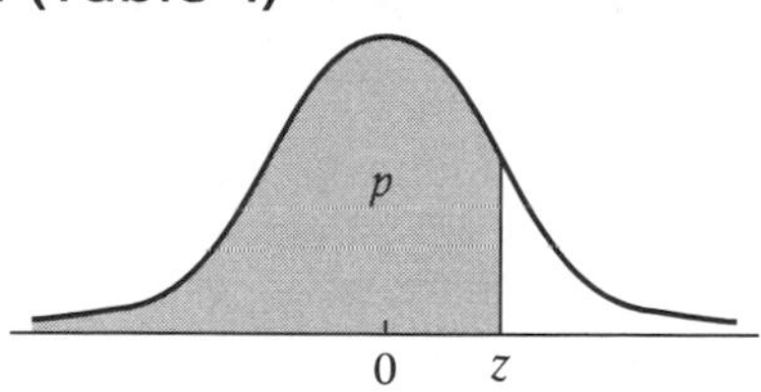

p	**0.00**	**0.01**	**0.02**	**0.03**	**0.04**	**0.05**	**0.06**	**0.07**	**0.08**	**0.09**	p
0.5	0.0000	0.0251	0.0502	0.0753	0.1004	0.1257	0.1510	0.1764	0.2019	0.2275	**0.5**
0.6	0.2533	0.2793	0.3055	0.3319	0.3585	0.3853	0.4125	0.4399	0.4677	0.4958	**0.6**
0.7	0.5244	0.5534	0.5828	0.6128	0.6433	0.6745	0.7063	0.7388	0.7722	0.8064	**0.7**
0.8	0.8416	0.8779	0.9154	0.9542	0.9945	1.0364	1.0803	1.1264	1.1750	1.2265	**0.8**
0.9	1.2816	1.3408	1.4051	1.4758	1.5548	1.6449	1.7507	1.8808	2.0537	2.3263	**0.9**

p	**0.000**	**0.001**	**0.002**	**0.003**	**0.004**	**0.005**	**0.006**	**0.007**	**0.008**	**0.009**	p
0.95	1.6449	1.6546	1.6646	1.6747	1.6849	1.6954	1.7060	1.7169	1.7279	1.7392	**0.95**
0.96	1.7507	1.7624	1.7744	1.7866	1.7991	1.8119	1.8250	1.8384	1.8522	1.8663	**0.96**
0.97	1.8808	1.8957	1.9110	1.9268	1.9431	1.9600	1.9774	1.9954	2.0141	2.0335	**0.97**
0.98	2.0537	2.0749	2.0969	2.1201	2.1444	2.1701	2.1973	2.2262	2.2571	2.2904	**0.98**
0.99	2.3263	2.3656	2.4089	2.4573	2.5121	2.5758	2.6521	2.7478	2.8782	3.0902	**0.99**

PERCENTAGE POINTS OF THE STUDENT'S t-DISTRIBUTION (Table 5)

The table gives the values of z satisfying $P(X \leqslant x) = p$, where X is a random variable having the Student's t-distribution with ν degrees of freedom.

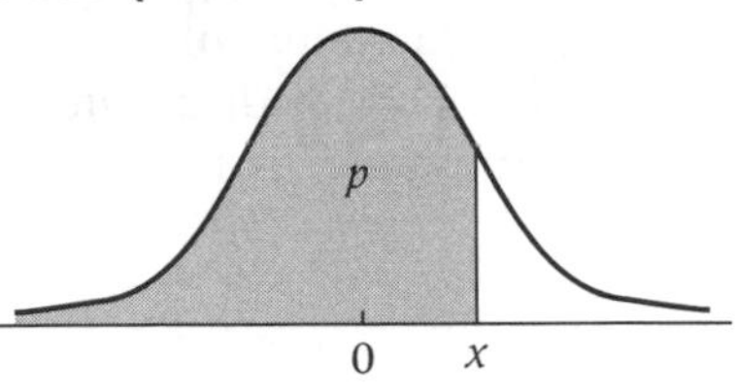

p / ν	0.9	0.95	0.975	0.99	0.995
1	3.078	6.314	12.706	31.821	63.657
2	1.886	2.920	4.303	6.965	9.925
3	1.638	2.353	3.182	4.541	5.841
4	1.533	2.132	2.776	3.747	4.604
5	1.476	2.015	2.571	3.365	4.032
6	1.440	1.943	2.447	3.143	3.707
7	1.415	1.895	2.365	2.998	3.499
8	1.397	1.860	2.306	2.896	3.355
9	1.383	1.833	2.262	2.821	3.250
10	1.372	1.812	2.228	2.764	3.169
11	1.363	1.796	2.201	2.718	3.106
12	1.356	1.782	2.179	2.681	3.055
13	1.350	1.771	2.160	2.650	3.012
14	1.345	1.761	2.145	2.624	2.977
15	1.341	1.753	2.131	2.602	2.947
16	1.337	1.746	2.121	2.583	2.921
17	1.333	1.740	2.110	2.567	2.898
18	1.330	1.734	2.101	2.552	2.878
19	1.328	1.729	2.093	2.539	2.861
20	1.325	1.725	2.086	2.528	2.845
21	1.323	1.721	2.080	2.518	2.831
22	1.321	1.717	2.074	2.508	2.819
23	1.319	1.714	2.069	2.500	2.807
24	1.318	1.711	2.064	2.492	2.797
25	1.316	1.708	2.060	2.485	2.787
26	1.315	1.706	2.056	2.479	2.779
27	1.314	1.703	2.052	2.473	2.771
28	1.313	1.701	2.048	2.467	2.763
29	1.311	1.699	2.045	2.462	2.756
30	1.310	1.697	2.042	2.457	2.750
31	1.309	1.696	2.040	2.453	2.744
32	1.309	1.694	2.037	2.449	2.738
33	1.308	1.692	2.035	2.445	2.733
34	1.307	1.691	2.032	2.441	2.728
35	1.306	1.690	2.030	2.438	2.724
36	1.306	1.688	2.028	2.434	2.719
37	1.305	1.687	2.026	2.431	2.715
38	1.304	1.686	2.024	2.429	2.712
39	1.304	1.685	2.023	2.426	2.708
40	1.303	1.684	2.021	2.423	2.704
45	1.301	1.679	2.014	2.412	2.690
50	1.299	1.676	2.009	2.403	2.678
55	1.297	1.673	2.004	2.396	2.668
60	1.296	1.671	2.000	2.390	2.660
65	1.295	1.669	1.997	2.385	2.654
70	1.294	1.667	1.994	2.381	2.648
75	1.293	1.665	1.992	2.377	2.643
80	1.292	1.664	1.990	2.374	2.639
85	1.292	1.663	1.998	2.371	2.635
90	1.291	1.662	1.987	2.368	2.632
95	1.291	1.661	1.985	2.366	2.629
100	1.290	1.660	1.984	2.364	2.626
125	1.288	1.657	1.979	2.357	2.616
150	1.287	1.655	1.976	2.351	2.609
200	1.286	1.653	1.972	2.345	2.601
∞	1.282	1.645	1.960	2.326	2.576

PERCENTAGE POINTS OF THE χ^2 DISTRIBUTION (Table 6)

The table gives the values of x satisfying $P(X \leq x) = p$, where X is a random variable having the χ^2 distribution with ν degrees of freedom.

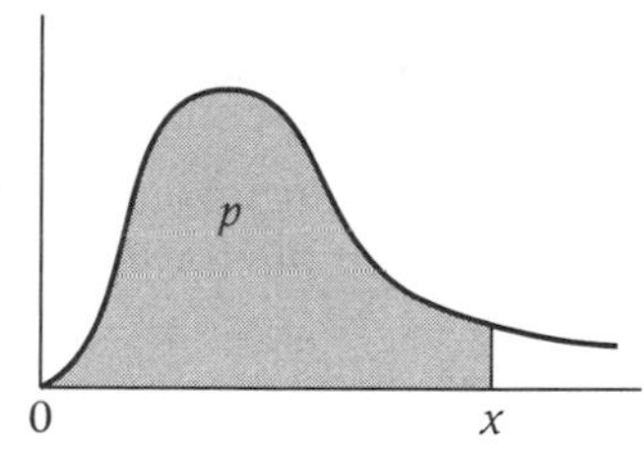

p / ν	0.005	0.01	0.025	0.05	0.1	0.9	0.95	0.975	0.99	0.995	p / ν
1	0.00004	0.0002	0.001	0.004	0.016	2.706	3.841	5.024	6.635	7.879	1
2	0.010	0.020	0.051	0.103	0.211	4.605	5.991	7.378	9.210	10.597	2
3	0.072	0.115	0.216	0.352	0.584	6.251	7.815	9.348	11.345	12.838	3
4	0.207	0.297	0.484	0.711	1.064	7.779	9.488	11.143	13.277	14.860	4
5	0.412	0.554	0.831	1.145	1.610	9.236	11.070	12.833	15.086	16.750	5
6	0.676	0.872	1.237	1.635	2.204	10.645	12.592	14.449	16.812	18.548	6
7	0.989	1.239	1.690	2.167	2.833	12.017	14.067	16.013	18.475	20.278	7
8	1.344	1.646	2.180	2.733	3.490	13.362	15.507	17.535	20.090	21.955	8
9	1.735	2.088	2.700	3.325	4.168	14.684	16.919	19.023	21.666	23.589	9
10	2.156	2.558	3.247	3.940	4.865	15.987	18.307	20.483	23.209	25.188	10
11	2.603	3.053	3.816	4.575	5.578	17.275	19.675	21.920	24.725	26.757	11
12	3.074	3.571	4.404	5.226	6.304	18.549	21.026	23.337	26.217	28.300	12
13	3.565	4.107	5.009	5.892	7.042	19.812	22.362	24.736	27.688	29.819	13
14	4.075	4.660	5.629	6.571	7.790	21.064	23.685	26.119	29.141	31.319	14
15	4.601	5.229	6.262	7.261	8.547	22.307	24.996	27.488	30.578	32.801	15
16	5.142	5.812	6.908	7.962	9.312	23.542	26.296	28.845	32.000	34.267	16
17	5.697	6.408	7.564	8.672	10.085	24.769	27.587	30.191	33.409	35.718	17
18	6.265	7.015	8.231	9.390	10.865	25.989	28.869	31.526	34.805	37.156	18
19	6.844	7.633	8.907	10.117	11.651	27.204	30.144	32.852	36.191	38.582	19
20	7.434	8.260	9.591	10.851	12.443	28.412	31.410	34.170	37.566	39.997	20
21	8.034	8.897	10.283	11.591	13.240	29.615	32.671	35.479	38.932	41.401	21
22	8.643	9.542	10.982	12.338	14.041	30.813	33.924	36.781	40.289	42.796	22
23	9.260	10.196	11.689	13.091	14.848	32.007	35.172	38.076	41.638	44.181	23
24	9.886	10.856	12.401	13.848	15.659	33.196	36.415	39.364	42.980	45.559	24
25	10.520	11.524	13.120	14.611	16.473	34.382	37.652	40.646	44.314	46.928	25
26	11.160	12.198	13.844	15.379	17.292	35.563	38.885	41.923	45.642	48.290	26
27	11.808	12.879	14.573	16.151	18.114	36.741	40.113	43.195	46.963	49.645	27
28	12.461	13.565	15.308	16.928	18.939	37.916	41.337	44.461	48.278	50.993	28
29	13.121	14.256	16.047	17.708	19.768	39.087	42.557	45.722	49.588	52.336	29
30	13.787	14.953	16.791	18.493	20.599	40.256	43.773	46.979	50.892	53.672	30
31	14.458	15.655	17.539	19.281	21.434	41.422	44.985	48.232	52.191	55.003	31
32	15.134	16.362	18.291	20.072	22.271	42.585	46.194	49.480	53.486	56.328	32
33	15.815	17.074	19.047	20.867	23.110	43.745	47.400	50.725	54.776	57.648	33
34	16.501	17.789	19.806	21.664	23.952	44.903	48.602	51.996	56.061	58.964	34
35	17.192	18.509	20.569	22.465	24.797	46.059	49.802	53.203	57.342	60.275	35
36	17.887	19.223	21.336	23.269	25.643	47.212	50.998	54.437	58.619	61.581	36
37	18.586	19.960	22.106	24.075	26.492	48.363	52.192	55.668	59.892	62.883	37
38	19.289	20.691	22.878	24.884	27.343	49.513	53.384	56.896	61.162	64.181	38
39	19.996	21.426	23.654	25.695	28.196	50.660	54.572	58.120	62.428	65.476	39
40	20.707	22.164	24.433	26.509	29.051	51.805	55.758	59.342	63.691	66.766	40
45	24.311	25.901	28.366	30.612	33.350	57.505	61.656	65.410	69.957	73.166	45
50	27.991	29.707	32.357	34.764	37.689	63.167	67.505	71.420	76.154	79.490	50
55	31.735	33.570	36.398	38.958	42.060	68.796	73.311	77.380	82.292	85.749	55
60	35.534	37.485	40.482	43.188	46.459	74.397	79.082	83.298	88.379	91.952	60
65	39.383	41.444	44.603	47.450	50.883	79.973	84.821	89.177	94.422	98.105	65
70	43.275	45.442	48.758	51.739	55.329	85.527	90.531	95.023	100.425	104.215	70
75	47.206	49.475	52.942	56.054	59.795	91.061	96.217	100.839	106.393	110.286	75
80	51.172	53.540	57.153	60.391	64.278	96.578	101.879	106.629	112.329	116.321	80
85	55.170	57.634	61.389	64.749	68.777	102.079	107.522	112.393	118.236	122.325	85
90	59.196	61.754	65.647	69.126	73.291	107.565	113.145	118.136	124.116	128.299	90
95	63.250	65.898	69.925	73.520	77.818	113.038	118.752	123.858	129.973	134.247	95
100	67.328	70.065	74.222	77.929	82.358	118.498	124.342	129.561	135.807	140.169	100

Formulae

This section lists formulae which relate to the Statistics module MS2, and which are included in the AQA formulae booklet.

Discrete distributions

For a discrete random variable X taking values x_i with probabilities p_i

Expectation (mean): $\mathrm{E}(X) = \mu = \sum x_i p_i$

Variance:
$$\mathrm{Var}(X) = \sigma^2 = \sum (x_i - \mu)^2 p_i = \sum x_i^2 p_i - \mu^2 = \mathrm{E}(X^2) - \mu^2$$

For a function $\mathrm{g}(X)$: $\mathrm{E}(\mathrm{g}(X)) = \sum \mathrm{g}(x_i) p_i$

Distribution of X	$\mathrm{P}(X = x)$	Mean	Variance
Poisson $\mathrm{Po}(\lambda)$	$\mathrm{e}^{-\lambda}\dfrac{\lambda^x}{x!}$	λ	λ

Continuous distributions

For a continuous random variable X having probability density function $\mathrm{f}(x)$

Expectation (mean): $\mathrm{E}(X) = \mu = \int x\,\mathrm{f}(x)\,\mathrm{d}x$

Variance:
$$\mathrm{Var}(X) = \sigma^2 = \int (x - \mu)^2\,\mathrm{f}(x)\,\mathrm{d}x = \int x^2\,\mathrm{f}(x)\,\mathrm{d}x - \mu^2 = \mathrm{E}(X^2) - \mu^2$$

For a function $\mathrm{g}(X)$: $\mathrm{E}(\mathrm{g}(X)) = \int \mathrm{g}(x)\mathrm{f}(x)\,\mathrm{d}x$

Cumulative distribution function: $\mathrm{F}(x) = \mathrm{P}(X \leqslant x) = \int_{-\infty}^{x} \mathrm{f}(t)\,\mathrm{d}t$

Distribution of X	Probability density function	Mean	Variance
Uniform (Rectangular) on $[a, b]$	$\dfrac{1}{b - a}$	$\frac{1}{2}(a + b)$	$\frac{1}{12}(b - a)^2$
Normal $\mathrm{N}(\mu, \sigma^2)$	$\dfrac{1}{\sigma\sqrt{2\pi}}\mathrm{e}^{-\frac{1}{2}\left(\frac{x-\mu}{\sigma}\right)^2}$	μ	σ^2

Distribution-free (non-parametric) tests

Contingency table tests: $\sum \frac{(O_i - t_i)^2}{t_i}$ is approximately distributed as χ^2

Sampling distributions

For a random sample $X_1, X_2, \ldots, X_n$ of n independent observations from a distribution having mean μ and variance σ^2

$\overline{X}$ is an unbiased estimator of μ, with $\text{Var}(\overline{X}) = \frac{\sigma^2}{n}$

S^2 is an unbiased estimator of σ^2, where $S^2 = \frac{\sum(X_i - \overline{X})^2}{n - 1}$

For a random sample of n observations from $N(\mu, \sigma^2)$

$$\frac{\overline{X} - \mu}{\frac{\sigma}{\sqrt{n}}} \sim N(0, 1)$$

$$\frac{\overline{X} - \mu}{\frac{S}{\sqrt{n}}} \sim t_{n-1}$$

Candidates should learn the following formula, which are **not** included in the AQA formulae booklet, but which may be required to answer questions:

$E(aX + b) = aE(X) + b$ and $\text{Var}(aX + b) = a^2\text{Var}(X)$

$P(a < X < b) = \int_a^b f(x)\,dx$

$P(\text{Type I error}) = P(\text{reject } H_0 \mid H_0 \text{ true})$ and $P(\text{Type II error}) = P(\text{accept } H_0 \mid H_0 \text{ false})$

$E_{ij} = \frac{R_i \times C_j}{T}$ and $v = (\text{rows} - 1)(\text{columns} - 1)$

Yates' correction (for 2×2 table) is $\chi^2 = \sum \frac{(|O_i - E_i| - 0.5)^2}{E_i}$

Notation

X, Y, R, etc.	random variables
x, y, r, etc.	values of the random variables X, Y, R, etc.
$x_1, x_2, \ldots$	observations
$f_1, f_2, \ldots$	frequencies with which the observations $x_1, x_2, \ldots$ occur
$p(x)$	probability function $\mathrm{P}(X = x)$ of the discrete random variable X
$p_1, p_2, \ldots$	probabilities of the values $x_1, x_2, \ldots$ of the discrete random variable X
$\mathrm{f}(x), \mathrm{g}(x), \ldots$	the value of the probability density function of the continuous random variable X.
$\mathrm{F}(x), \mathrm{G}(x), \ldots$	the value of the (cumulative) distribution function $\mathrm{P}(X \leqslant x)$ of the continuous random variable X
$\mathrm{E}(X)$	expectation of the random variable X
$\mathrm{E}[\mathrm{g}(X)]$	expectation of $\mathrm{g}(X)$
$\mathrm{Var}\,(X)$	variance of the random variable X
$\mathrm{N}(\mu, \sigma^2)$	normal distribution with a mean of μ and a variance σ^2
μ	population mean
σ^2	population variance
σ	population standard deviation
$\bar{x}$	sample mean
s^2	unbiased estimate of population variance from a sample, $s^2 = \dfrac{1}{n-1}\sum(x_i - \bar{x})^2$
z	value of the standardised normal variable with distribution $\mathrm{N}(0, 1)$
$\Phi(z)$	corresponding (cumulative) distribution function

Index